IMAGES
of America

HIGHLANDLAKE AND MEAD

On the Cover: This January 25, 1907, photograph commemorates the arrival of a group of Longmont Boosters, sponsored by the Denver & Great Western Railway, to the town of Mead. Greeting the train is the Mead Brass Band, which then escorted the dignitaries down Main Street while playing lively tunes. Paul Mead, founder of Mead, is third from the left, with his cornet tucked tightly under his arm. Paul and his wife, Ariet Palmer Mead, carefully drew up plans for the new town, naming the streets after family members and the town itself after Paul's father, Dr. Martin Luther Mead, MD. (Longmont Museum and Cultural Center.)

IMAGES
of America

HIGHLANDLAKE AND MEAD

Pauli Driver Smith

ARCADIA
PUBLISHING

ISBN 978-0-7385-9601-3

Published by Arcadia Publishing
Charleston, South Carolina

Printed in the United States of America

Library of Congress Control Number: 2012953038

For all general information, please contact Arcadia Publishing:
Telephone 843-853-2070
Fax 843-853-0044
E-mail sales@arcadiapublishing.com
For customer service and orders:
Toll-Free 1-888-313-2665

Visit us on the Internet at www.arcadiapublishing.com

This book is dedicated to Mary Mead Jensen, daughter, granddaughter, and cousin of Highlandlake and Mead Pioneers. Born at the family Highlandlake homestead, she lived nearly all of her 97 years in the home where she was born. She loved and served her community well.

Contents

Acknowledgments

In grateful acknowledgement to the many generous people who were so willing to allow me to scan their precious family photographs over the past 27 years, this book would not be possible without you. I especially want to thank the late Mary Mead Jensen and her family. Mary was so excited about the possibility of this book and could not wait to see it in print. Sadly, she passed away before she could, but her encouragement and interesting stories about early Highlandlake and Mead are what helped me fall in love with this area.

Special thanks go to the Historic Highlandlake, Inc., Board of Directors, who encouraged and supported me through the book-writing process. Lawrence Jensen especially helped locate many of the Mead family photographs that I originally scanned two decades ago but, for the purposes of this book, needed to rescan.

Grateful thanks go to Erik Mason, archivist at the Longmont Museum and Cultural Center. I surely tested his patience at times by my frequent requests for yet more local information and obscure photographs.

Finally, thanks go to Colorado East Bank and Trust and to Mead Family Dentistry for helping to cover the cost of obtaining scans and usage rights for photographs obtained from the collections of the Denver Public Library and Longmont Museum and Cultural Center.

INTRODUCTION

Unlike popular Hollywood images of the Wild West where gunslingers and desperadoes routinely tear up a town and the surrounding countryside, the majority of residents in Colorado's prairie villages and towns were educated, law-abiding, and hardworking folks. Some people came for their health, some sought free land, and a few, after failing to find their fortunes in the goldfields, traded in a miner's pick for a farmer's plow. They were entrepreneurs, farmers, ranchers, schoolteachers, preachers, doctors, nurses, immigrants, miners, and field-workers. While their motives for coming to the Highlandlake and Mead area were for different reasons, they all had the common goal of a better life for themselves and their families.

Highlandlake was progressive on many fronts. Most of the early residents were well educated and believed strongly in building their community. Within a few years, they established a school, small lending library, and a church. They also formed three literary/elocution societies (one each for the men, women, and children) and three bands, including juvenile, cornet, and brass bands. Everyone was actively involved in school and church activities. Clubs for all ages rapidly followed, such as the Women's Pleasure Club (an educational and friendship club for women), the Highlandlake Helpers (young men), and swimming, tennis, and croquet clubs. Just about the only club proposed that did not succeed was a cooking club that failed due to lack of interest.

Besides clubs and organizations, Highlandlake was known for its strong women. Early on, they embraced women's suffrage and the Women's Temperance Christian Union. In 1893, the women threatened to go on strike if their menfolk did not vote for women's suffrage. Suffrage passed, making Colorado the second state to give women the vote. The next week's issue of the *Longmont Ledger* duly reported that apparently the men voted as their women wished, as all was peaceful in Highlandlake. In 1894, church members hired a woman for their pastor. A year later, in 1895, Highlandlake elected one of the first all-women school boards in Colorado, beating Denver out by several years.

When building a town, both men and women realized the necessity of working closely together. Paul and Ariet Palmer Mead each worked on the town plans, including inserting a "dry clause" on all deeds to lots located on their land. More than one husband and wife team, such as Will and Pearl Markham, served on school and church boards together. Still, when it came to running a town, it was the men who stood in the forefront. It would take decades before a woman served on the Mead Board of Trustees and almost 90 years before a woman became the town's mayor.

After electing a mayor and board of trustees, one of the first orders of business was to hire H. Wilson as the first police magistrate, for $26 a year. By 1910, his wages were raised to $10 a month. Increased wages reflected the growing town's increased wealth, rather than the incidence of crime. Outside of the occasional bank or store robbery committed by outsiders, serious crimes were rare. Up until the early 1960s, most sojourns in the Mead jail were brief and for minor infractions, such as public drunkenness or for driving over Mead's rigidly enforced 10-miles-per-hour speed limit (eight around corners).

Much of the town's social agenda revolved around the weekends. Saturday was the day that area farmers brought in their goods to trade or sell and buy supplies. Taking advantage of the weekly influx of people, medicine shows, circuses, and other forms of traveling shows frequently entertained willing audiences. Dances were held year-round, first at the Rose Hall, owned by Charles A. Rose, and later at the Farmer's Union Hall, where Saturday matinees were also featured. During the 1930s and 1940s, summer evenings often included a movie projected onto the broad, white, exterior walls of Clark's Lumber.

Each community offered complementary and sometimes competing activities. Highlandlake had camping, swimming, fishing, boating, picnicking, and tennis. Mead, with its larger population, offered movies, bowling, and community fairs. Both had nine-man baseball teams and other sports, musical concerts, plays, and dances. Churches and schools in both communities offered opportunities for people to get together and nurture friendships.

As Mead entered the 1920s, a building boom was on. Crops and livestock were selling at record prices, and businesses thrived. With the stock market crash of 1929 coupled with severe drought, all of that changed. Crop and livestock prices plummeted. Within a few years, both banks failed and, with them, many of the businesses. Farms were foreclosed on in record numbers, forcing people to leave in search of work. Adding to the loss was that once the Depression was over, cars were making it convenient to drive to neighboring cities, such as Longmont, for supplies and services. A small town like Mead could no longer compete. A few businesses held on, but for the most part, Mead, like so many other small towns, settled into a relative obscurity.

It took until a housing boom in the 1990s to wake Mead up from its gentle slumber. With new housing subdivisions came more families, and with families came a need for improved roads and amenities. In a span of only a few years, the town's population grew from around 600 to over 2,000, causing it at one time in the 1990s to be named the fastest-growing town in Colorado. In the years since, growth has slowed but stayed steady. The 2010 Census shows the town's population at 3,405.

Today, Mead's advantageous location on both sides of Interstate 25, along with attractive housing prices and rural, small-town ambiance, is once again drawing people to this area. The reasons for settling in Mead today differ widely from those of the early pioneers. However, one thing is certain that no matter how you find your way to Highlandlake and Mead, or what you do once you arrive, this area calls you to stay and make your home in what is still known as the "Beautiful Highlandlake Country."

One

The Beautiful Highlandlake Country

On June 3, 1871, Lorin C. Mead, a recent arrival to the newly established Chicago-Colorado Colony (now Longmont), boarded Dave Baumert's stagecoach to Greeley. About eight miles into the trip, Mead noticed what he later described as a "singular, circular depression about 20 feet deep and 100 rods in diameter." In the middle of this muddy, spring-fed prairie pothole stood an antelope calmly drinking, seemingly unconcerned by the passing stagecoach. Impressed, Mead made a mental note of the location. Several months later, he staked his claim, thus beginning the story of Highlandlake.

In 1873, Mead added more land and finished the remainder of the Highland Ditch, bringing additional water storage, not only for his crops but also for hundreds of acres below the lake. Once a steady supply of irrigation water was established, others quickly took up homesteads in what became known as the Highlandlake District. In 1880, Lorin C. Mead and his friend George Davis organized a church. By 1883, the area was booming. Houses were going up everywhere, businesses opened, the schoolhouse added another two rooms to accommodate the growing number of students, and a community cemetery was established. As fame of the "Beautiful Highlandlake Country" spread, visitors from as far away as Denver and Colorado Springs flocked to its shores to enjoy camping, swimming, fishing, boating, and spectacular views of the Rocky Mountains.

By 1903, Highlandlake boasted over 200 permanent residents with a transient population swelling the numbers during summer months when visitors rented campsites or stayed in the Gateley Hotel. An acetylene gas line supplied homes, businesses, and the church with lighting, and it was only a matter of time before the promised Great Western Railroad arrived. However, in the fall of 1905, hopes of a railroad station and beet dump were dashed after the railroad announced that it was rerouting its tracks a mile and a half southeast of Highlandlake. Within a few years, all of the Highlandlake businesses closed, most of the residents moved to Mead, and the village of Highlandlake faded quietly into history.

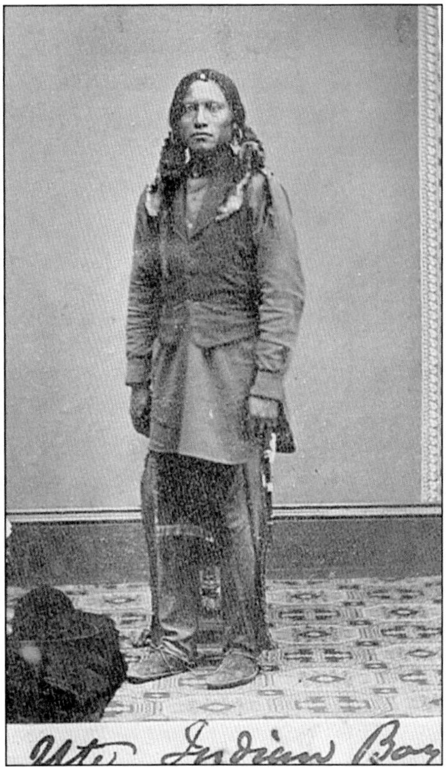

The first residents in the Highlandlake-Mead area were Native Americans. These included Cheyenne, Arapaho, Sioux, and Ute. When Isadore Terry True settled with her new husband in Highlandlake around 1873, she started collecting trading cards depicting members of different tribes. This card of a young Ute man is one of only two that survived the years. (Mary True-Farr collection, Historic Highlandlake.)

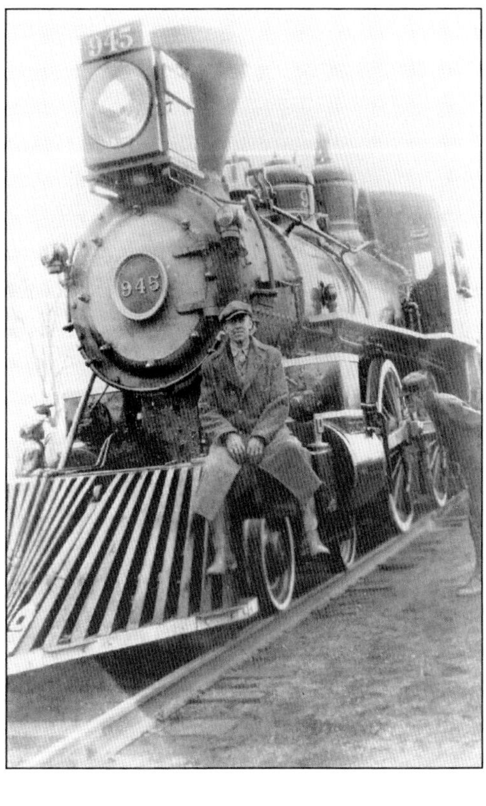

Until about 1875, early passengers first took this train to Erie and then rode a stage the rest of the distance to get to Highlandlake. In 1925, on its way to an exposition in California, the original engine stopped in Platteville for a brief layover. Carl True sits on the pilot while his father, Longmont and Highlandlake pioneer R.S. True, looks on. (Mary True-Farr collection, Historic Highlandlake.)

Lorin C. Mead was born in Vermont in 1832. He came to Colorado in 1871 and homesteaded 80 acres of land including the body of water he named Highland Lake. He later added more acreage. Mead founded the Highlandlake Church, served in the first state legislature in 1878, and served two terms as Weld County clerk and recorder. (Mary Mead Jensen family collection, Historic Highlandlake.)

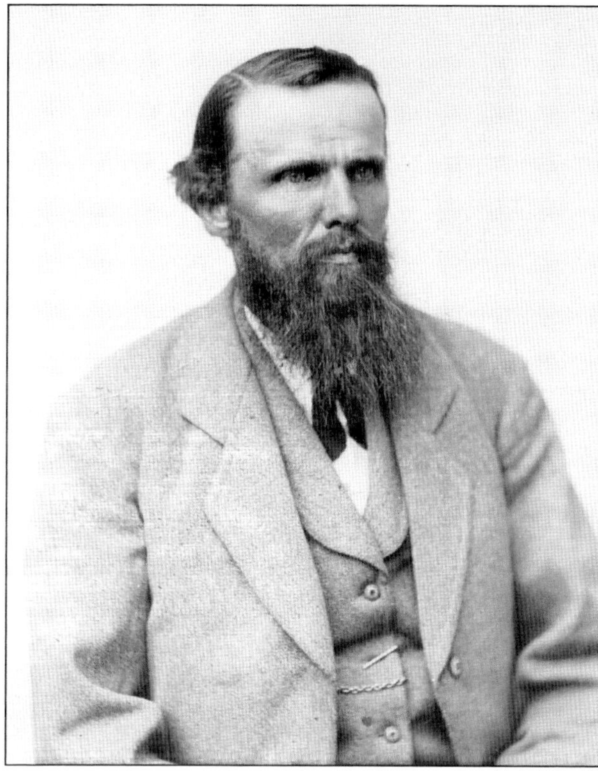

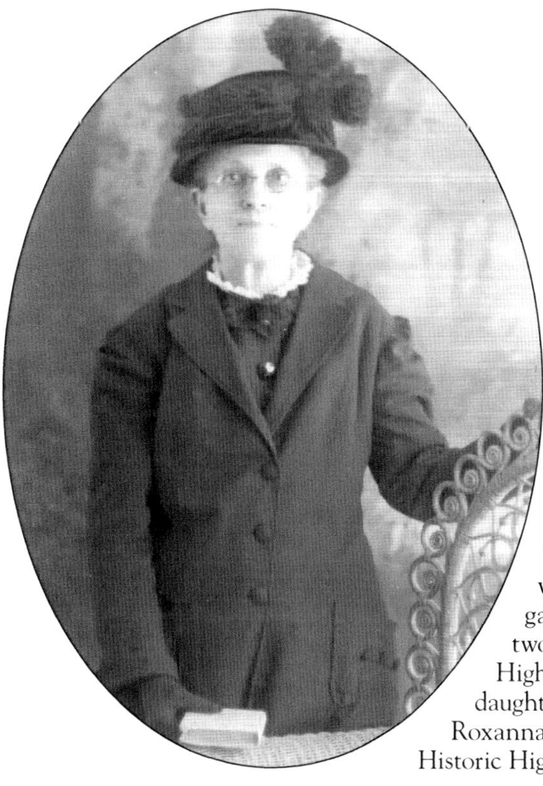

Elizabeth Sheldon Mead, affectionately called, "The Lady of the Lake," by her husband, Lorin C. Mead, came to Colorado with her husband in late May 1871. She gave Lorin two sons, Malcolm (born just two days after she moved to her new home in Highlandlake) and Rufus. She also raised Lorin's daughter Electa "Lettie" from his first marriage, to Roxanna Peat. (Mary Mead Jensen family collection, Historic Highlandlake.)

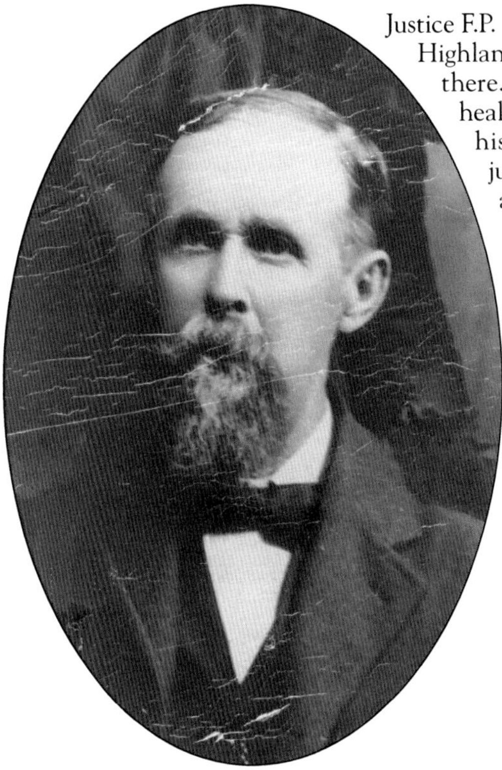

Justice F.P. (Frederick) Waite was the second settler in the Highlandlake District and the first to finish his house there. Waite came to Colorado about 1869 for his health. The cure worked, allowing him to live into his 80s. Waite was appointed the Highlandlake justice of the peace sometime around 1880–1883 and continued in that capacity until about 1910. (Longmont Museum and Cultural Center.)

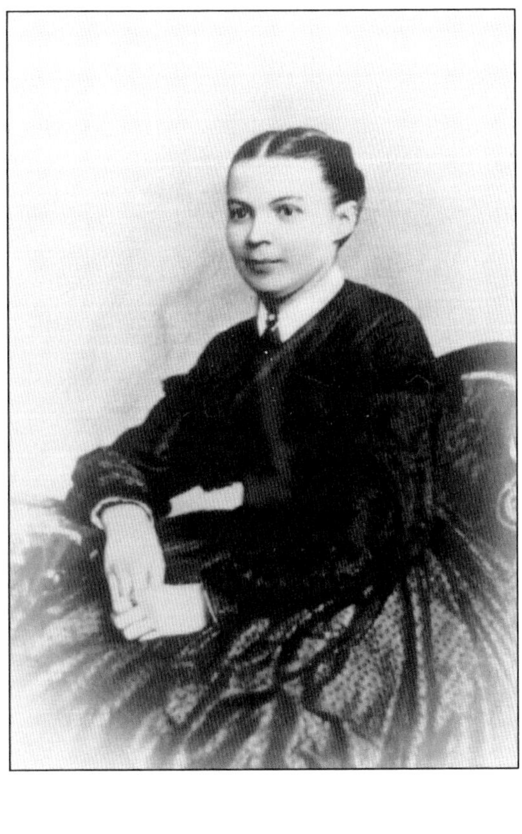

Hattie "Effie" Fowler Waite, the first wife of F.P. Waite, married her husband on October 4, 1871, in Denver. Their son Charlie was born a year later. They moved to Highlandlake in 1873. Tragically, on December 8, 1884, Charlie fell through the ice and drowned in Highland Lake. Effie, Charlie, and Effie's mother, Phoebe Fowler, are buried in the Highlandlake Pioneer Cemetery. (Betty Ann Newby collection, Historic Highlandlake.)

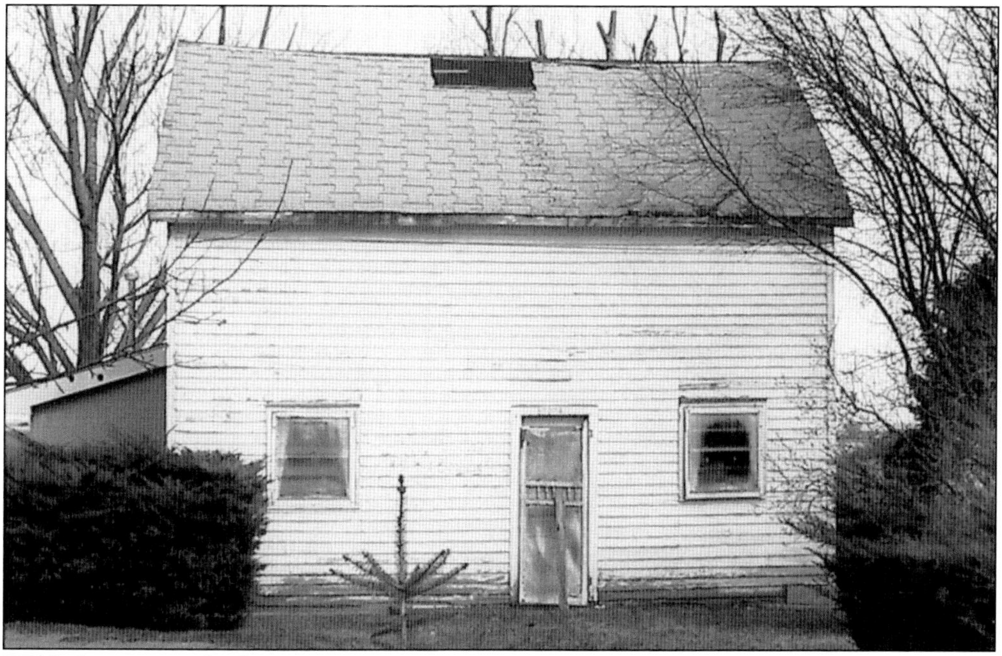

While he was not the first to file on land in Highlandlake, F.P. Waite was the first to finish his house, naming it The Ivies. In the fall of 1874, the first three families (Meads, Waites, and Pounds) living in what would eventually become Highlandlake celebrated Thanksgiving in his home. All three families later documented their memories of that day. (Pauli Driver Smith collection, Historic Highlandlake.)

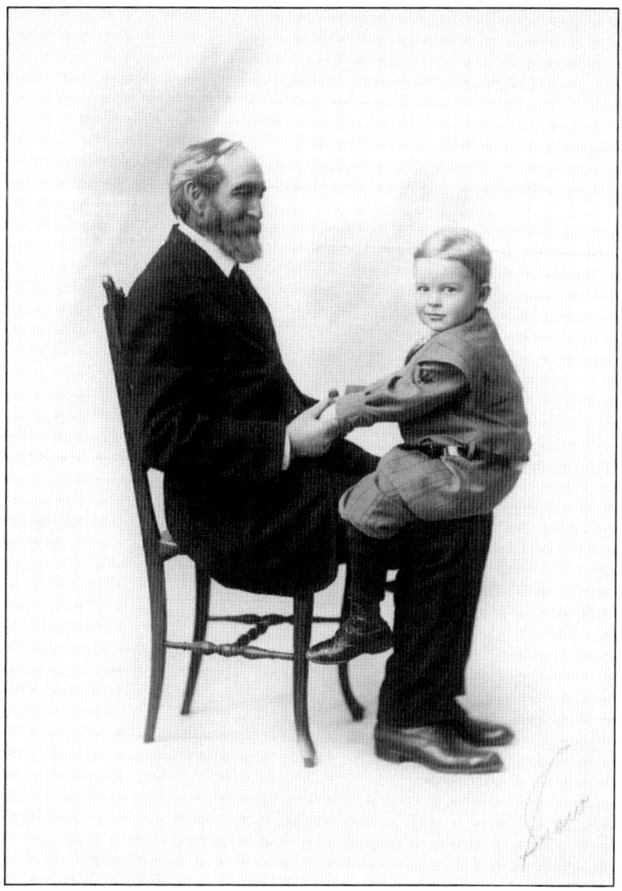

Justice F.P. Waite and grandson John Frederick Thompson, son of Waite's youngest daughter, Marietta Waite, and her first husband, Royal Thompson, pose for a delightful portrait around 1910 that was probably taken right about the time that Justice Waite moved to Boulder. (Betty Ann Newby/ John F. Thompson collection, Historic Highlandlake.)

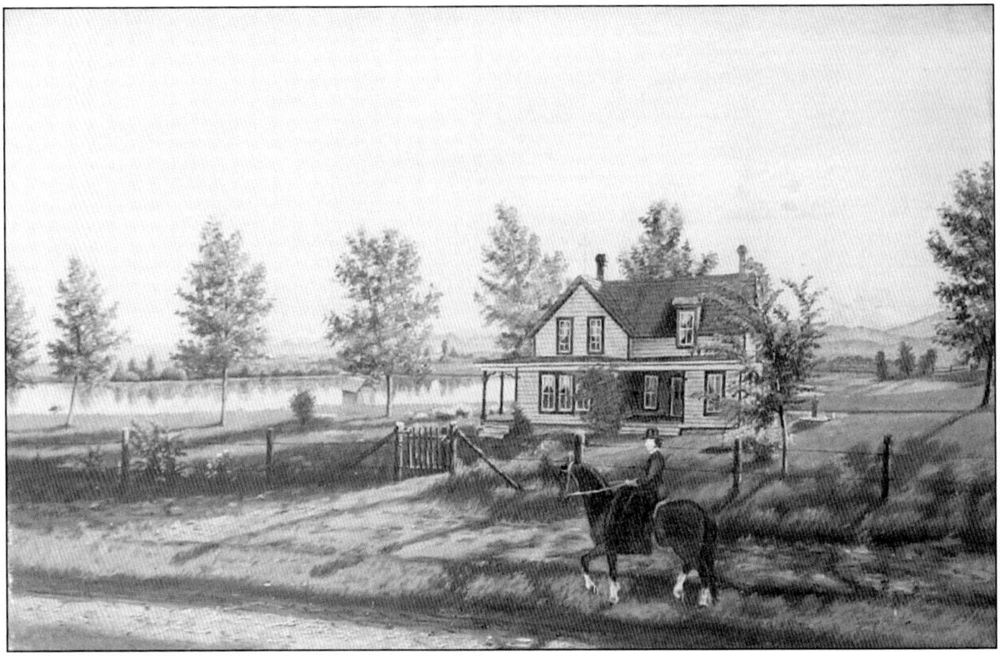

Shown above is a beautiful painting of Lorin C. Mead's home by an unidentified artist. It depicts the homestead around 1878–1879. Elizabeth Mead is riding sidesaddle on the horse in front of the home. The icehouse is just to the left of the house. The photograph below was taken about 1883–1884. The trees are taller and Elizabeth's eldest son, Malcolm "Mac" Mead, born in 1875, is showing off his modern Safety Bicycle. Before the days of pneumatic (inflated) wheels, to say that the ride was quite bumpy was an understatement. Note how the back wheel is larger than the front. This experimental style never really took off and disappeared from the bicycle scene within a few years. (Both, Mary Mead Jensen family collection, Historic Highlandlake.)

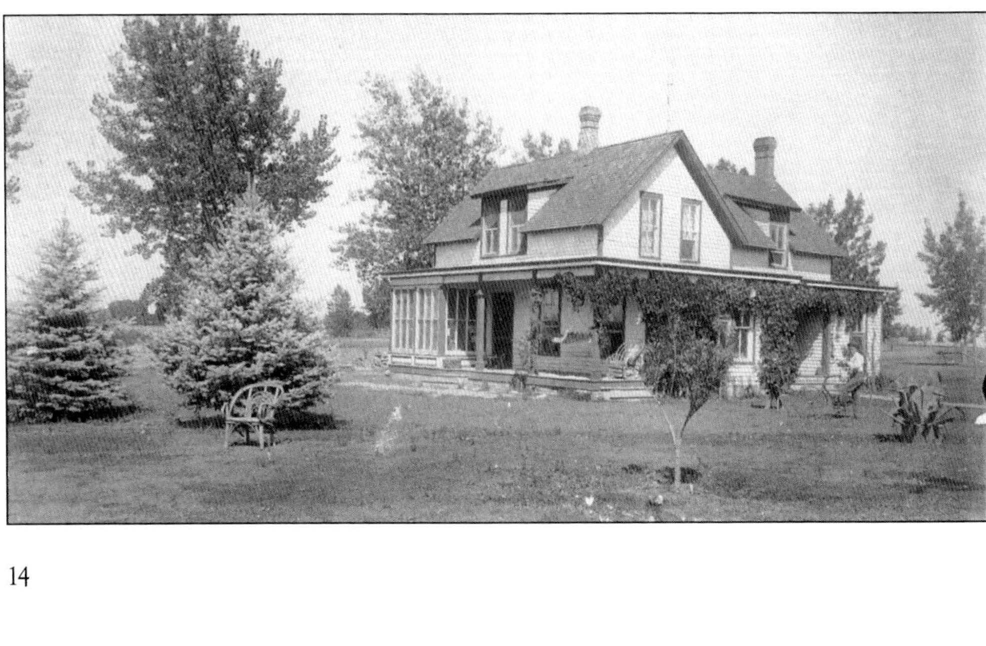

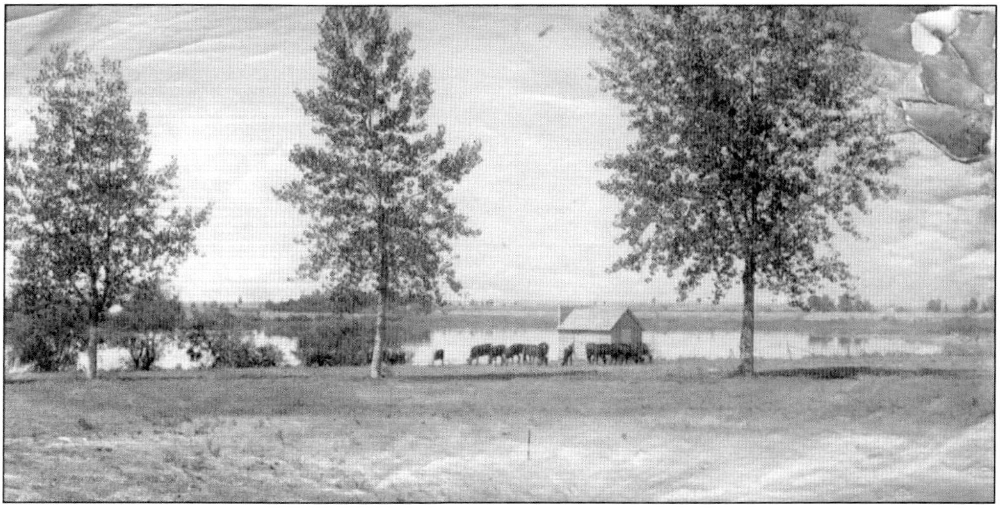

Residents harvested ice from the lake each winter and kept it packed in sawdust in the small, insulated building seen in this photograph. Ice harvesting began when the ice was at least 18 inches thick. Properly stored and rationed, the blocks of ice lasted well into the fall. (Mary Mead Jensen family collection, Historic Highlandlake.)

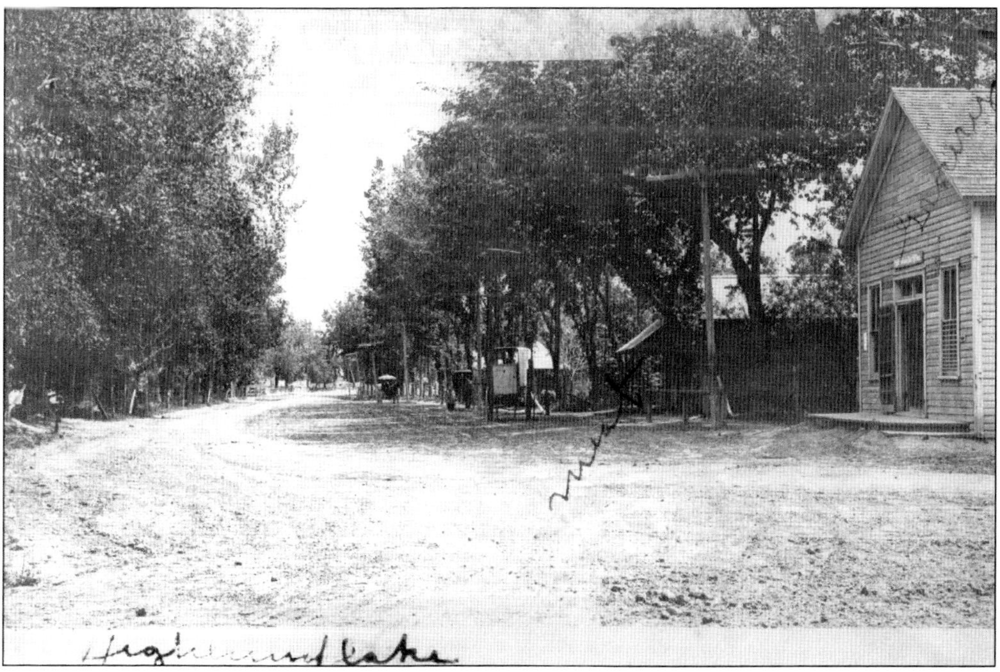

This real-photo postcard around 1906 shows some of the businesses originally located on the north end of Main Street. Richey's Grocery is first. Next is the bank, and then Hemingson's Meat Market. The barbershop, confectionery, fraternity hall, and M.S. Adams and Company were located around the corner. Other businesses included a livery stable, feed store, two blacksmiths, and a feed mill. (Mary Mead Jensen family collection, Historic Highlandlake.)

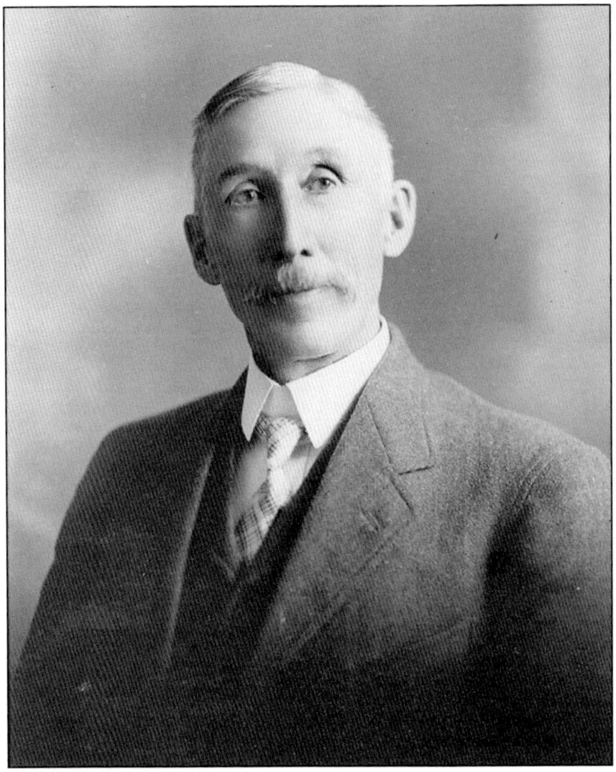

Rowland S. True followed his sweetheart, Isadore Terry, from Illinois to Colorado in 1871. He set up a jewelry store in Longmont and married Isadore the following year. They moved to Highlandlake sometime in late 1873 or early 1874 and, except for a few years in the late 1890s, lived there the rest of their lives. (Mary True-Farr collection, Historic Highlandlake.)

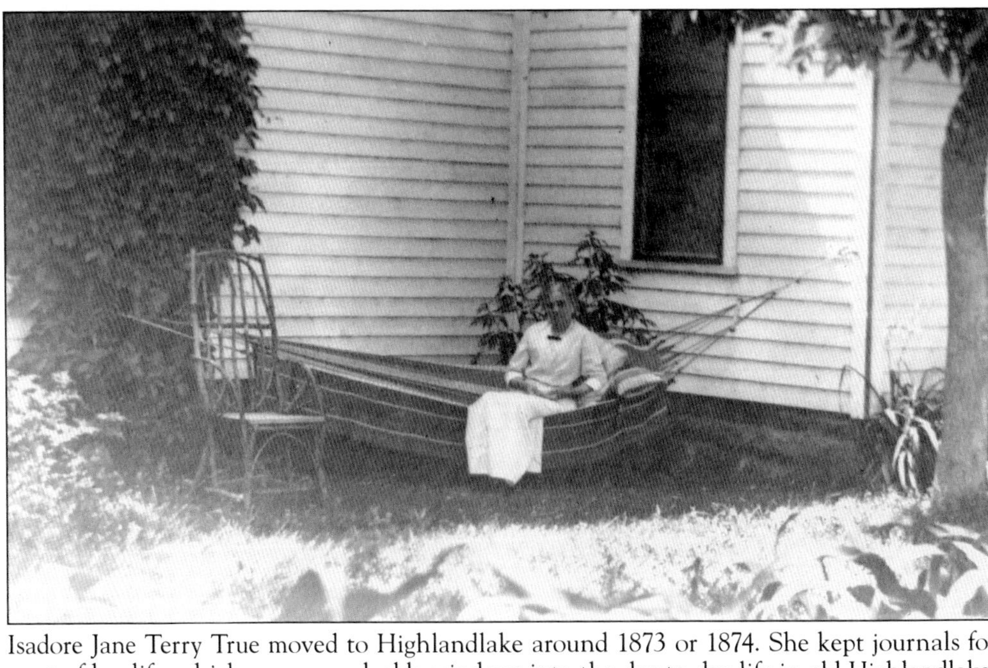

Isadore Jane Terry True moved to Highlandlake around 1873 or 1874. She kept journals for most of her life, which are now valuable windows into the day-to-day life in old Highlandlake. She is shown here, visiting her first home in Highlandlake in the summer of 1915. She died at age 66 the following November. The house is no longer standing. (Mary True-Farr collection, Historic Highlandlake.)

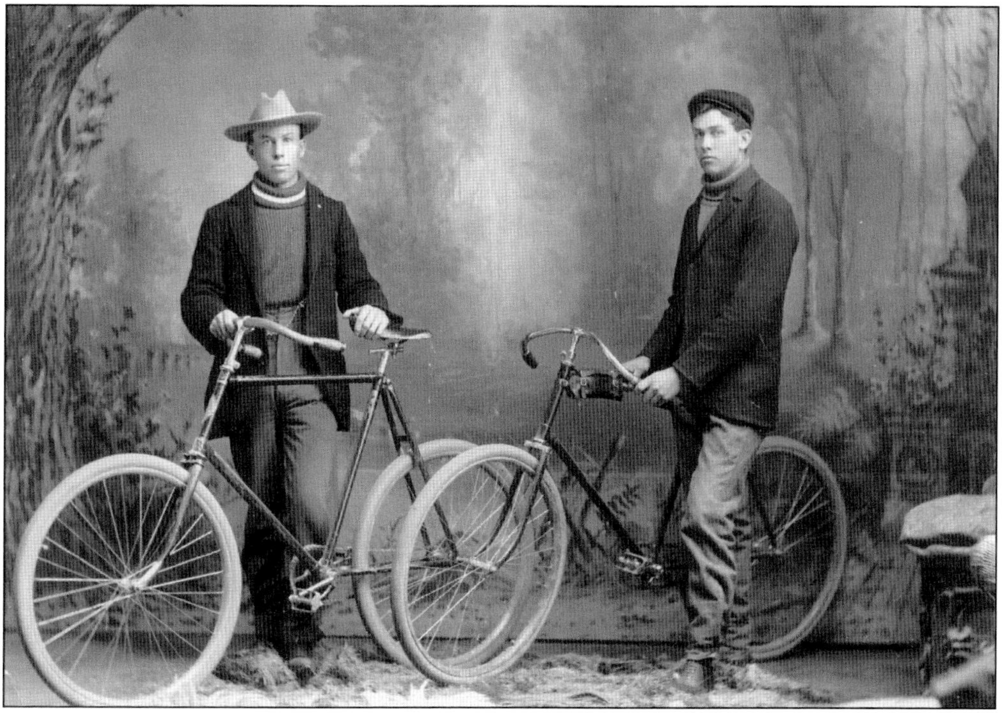

Ren (left) and Carlton "Carl" True, sons of R.S. and Isadore Terry True, show off their new wheels. The photograph was probably taken in a photography studio in Longmont around 1898. The brothers lived in the same home most of their lives, even after Carlton married, first to Stella Doughty and, then after her death, to Catherine Hildreth. (Mary Mead Jensen family collection, Historic Highlandlake.)

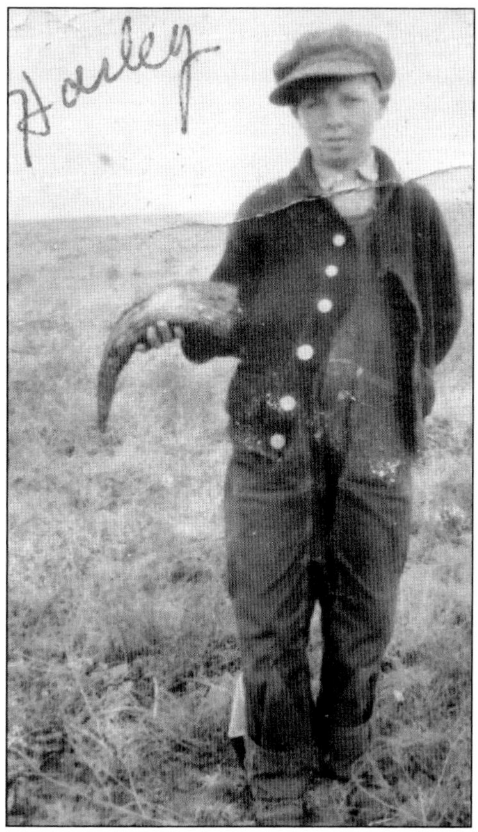

Harley Bruce Markham, son of Charles and Effie Evans Markham, was born in Highlandlake in 1904. When Harley was a boy, it was still possible for children to find buffalo horns and Indian arrowheads, whether they were digging a garden in the spring or helping to plow their father's fields. (Hester Markham collection, Historic Highlandlake.)

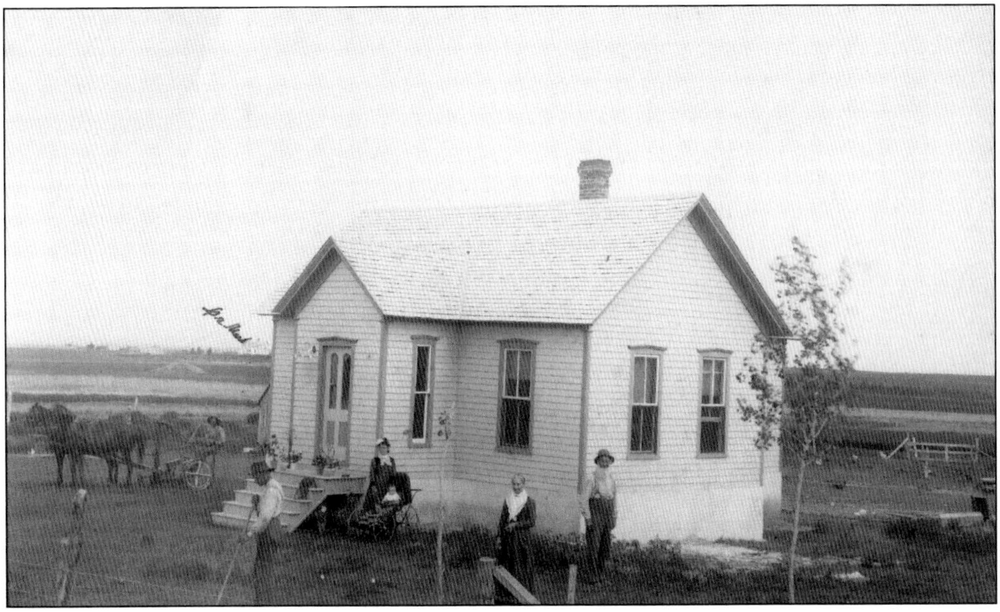

George Davis homesteaded the southeast corner of the lake in 1878. It took him almost five years to build his modest four-room home and satisfy his claim. Shown are, from left to right, son Charles K. (with team), son George Martin, Mrs. George Martin (Sarah) Davis with daughter Lilla Mae, Mrs. George (Mary) Davis, and George Davis Sr. Below, in 1887, the Davis family sold the property to the Highlandlake Church for a parsonage. In 1921, the property sold at auction to the Eckman family, who added the screened porch. The Eckmans operated a dairy farm and apple orchard and helped an uncle who raised bees. The sign on the tree proclaims, "Honey. We Produce it, We sell it." (Above, Longmont Museum and Cultural Center; below, Jean Eckman Nelson collection, Historic Highlandlake.)

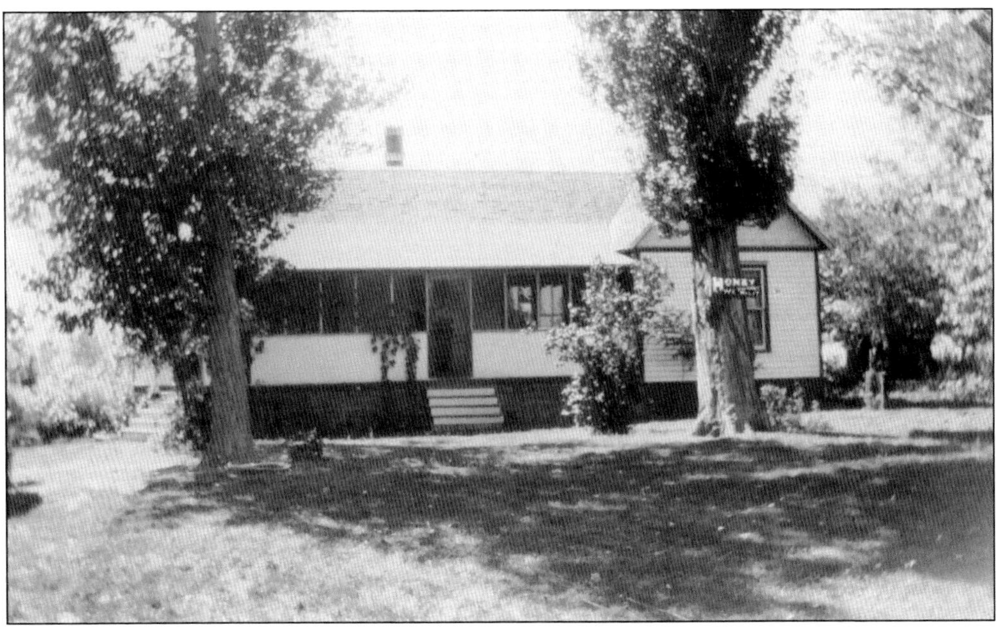

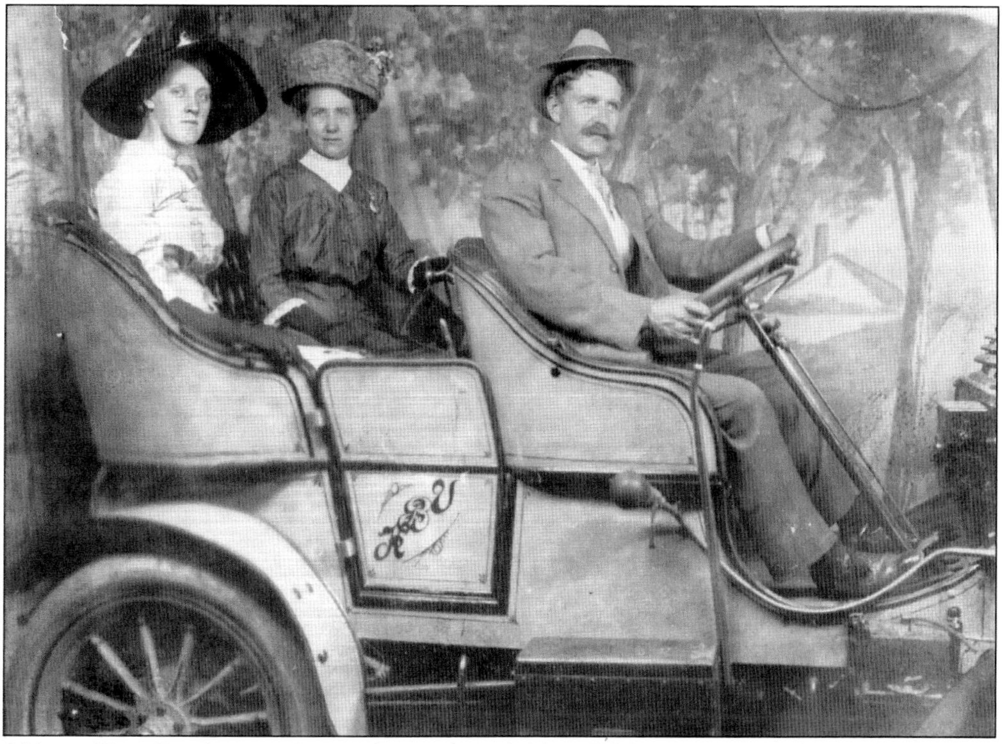

Siblings Elsie, Emely, Ivar, Otto, John, and Emil Olson/Olsberg immigrated to Colorado from Sweden in the early 1900s. Since money was scarce, John came over first. He worked in the mines until, one at a time, he could afford to bring the others to the United States. For a while, their last names changed frequently until each found a spelling he or she liked. All of the siblings except one eventually used Olson; John preferred Olsberg. Five of the siblings moved to Highlandlake, while Otto moved to Englewood. In this 1919 Denver studio portrait, shown above, sisters Elsie (left) and Emely and brother Otto chose this interesting photographer's prop to pose in. Shown below, on September 27, 1930, all six siblings got together for a family portrait. They are, from left to right, (first row) Emil Olson, Otto Olson, and John Olsberg; (second row) Emely Olson Eckman, Ivar Olson, and Elsie Olson Norman. (Both, Jean Eckman Nelson collection, Historic Highlandlake.)

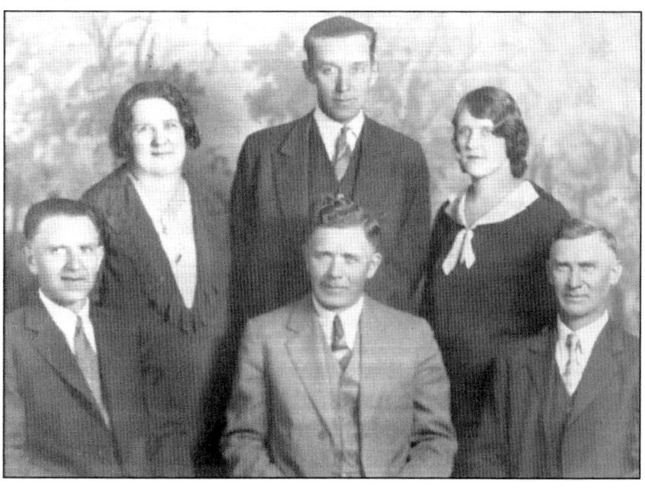

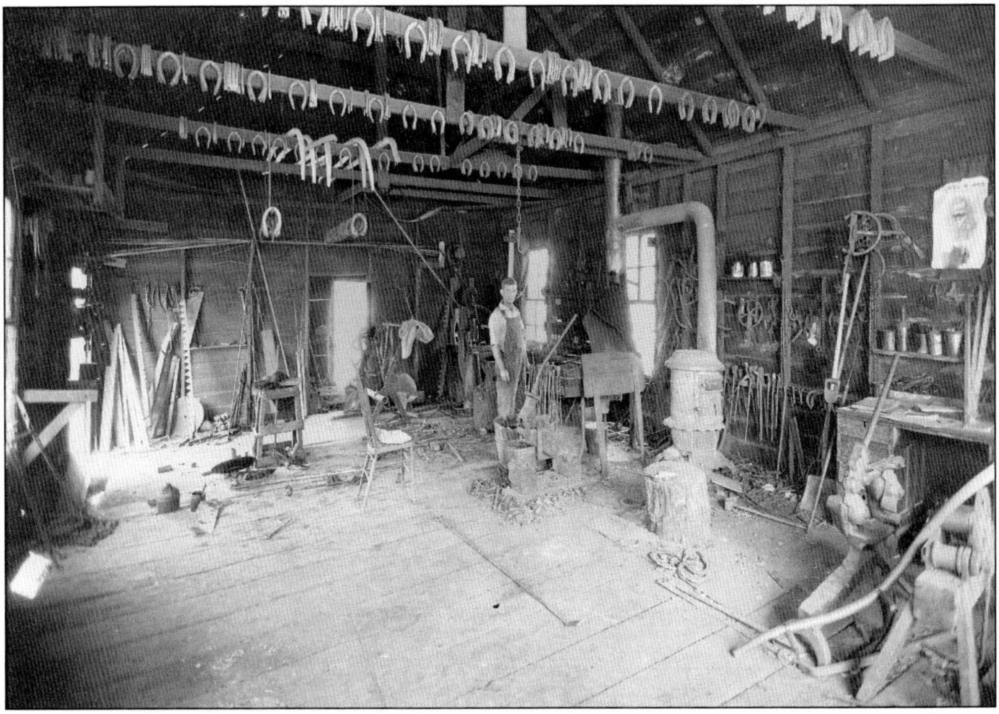

Harry Glover and Louis Eli Coates (brother of Forbes P. Coates) went into the livery and blacksmith business together sometime around 1905 or 1906. About the same time, a photographer named Purcell took these professional photographs. Their blacksmith shop and livery were located on the south side of Longs Peak Avenue (now Weld County Road 36) near Main Street (now Weld County Road 5). By 1910, Louis was divorced and co-owned a livery in Longmont with Roy Rhodes. The *Longmont Ledger* mentions a few years later that he is now in Mead, owning yet another livery. By 1930, Louis moved to Wyoming, then to Oregon, where he died in 1950. (Both, Mark French collection, Historic Highlandlake.)

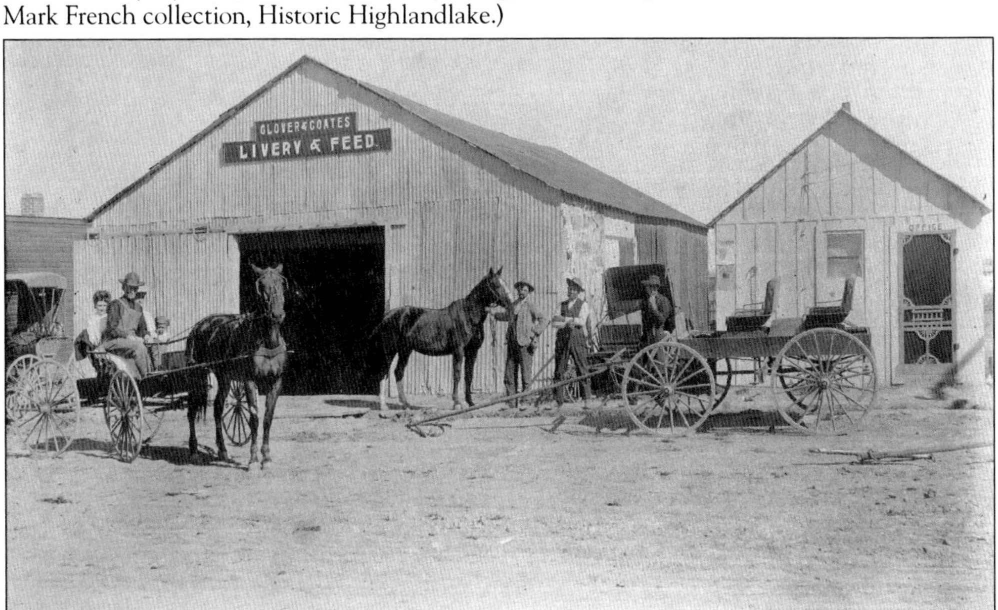

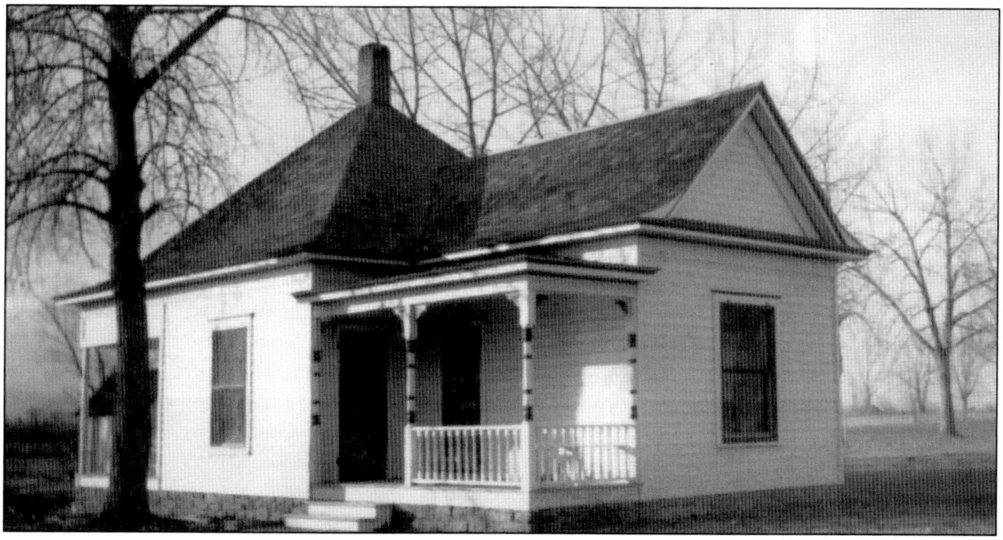

This house is sometimes called the old blacksmith's home, but it is unclear who actually built it. When Malcolm Mead, son of Lorin C. and Elizabeth Mead, and Florence Baker married in 1906, they made this house their first home. It is still standing and located just north of the café and concession stand at the lake. (Mary Mead Jensen family collection, Historic Highlandlake.)

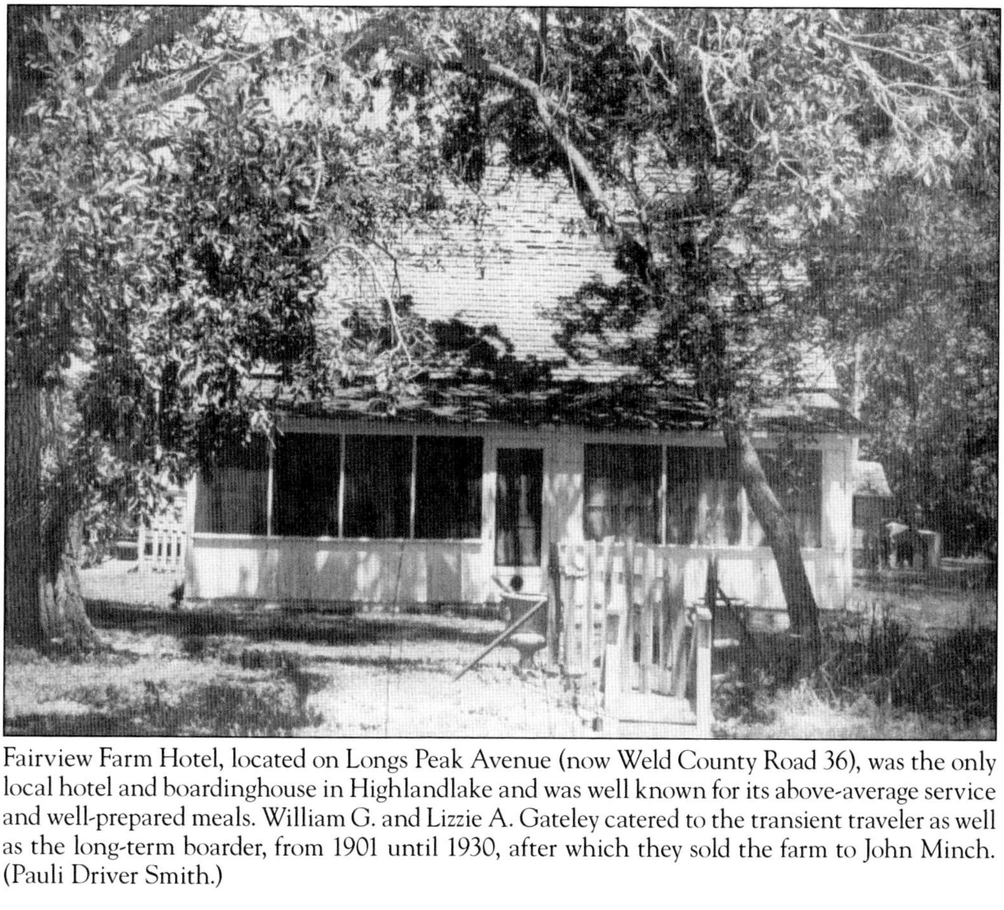

Fairview Farm Hotel, located on Longs Peak Avenue (now Weld County Road 36), was the only local hotel and boardinghouse in Highlandlake and was well known for its above-average service and well-prepared meals. William G. and Lizzie A. Gateley catered to the transient traveler as well as the long-term boarder, from 1901 until 1930, after which they sold the farm to John Minch. (Pauli Driver Smith.)

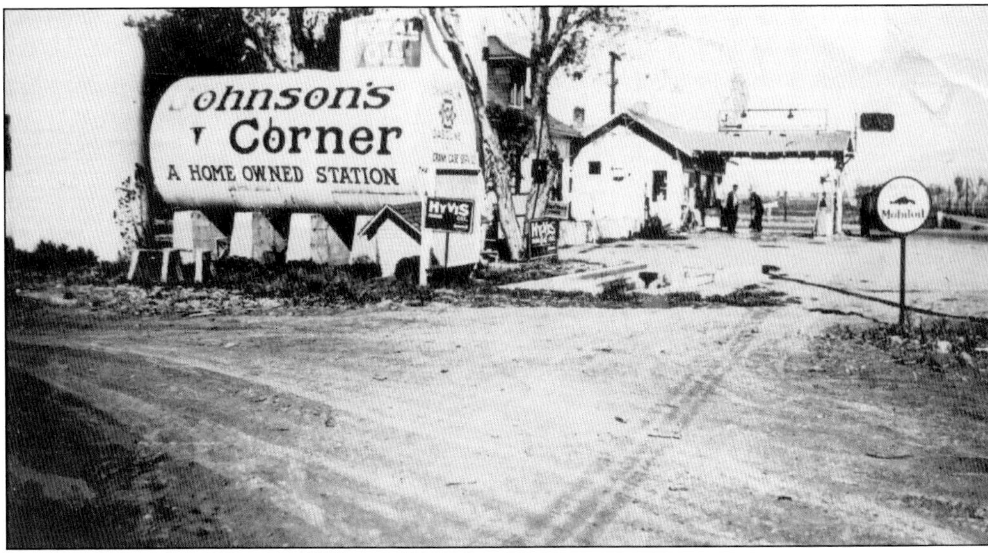

Long before Johnson's Corner in Loveland served its world-famous cinnamon rolls and even before the Johnson's Corner on South Main Street in Longmont served its first hamburger, there was a Johnson's Corner gas station located at what is now Seventeenth Avenue and Main Street in Longmont. Once the new station was built in 1937 on Longmont's South Main Street, the original building went up for sale. In 1941, Malcolm Mead purchased the old building and moved it to Highlandlake. The photograph above shows the structure in its original location. The photograph below is what it looked like just before it being lowered down on its new foundation. Mead turned the former gas station into a rental house. The house is still standing and lived in to this day. (Both, Mary Mead Jensen family collection, Historic Highlandlake.)

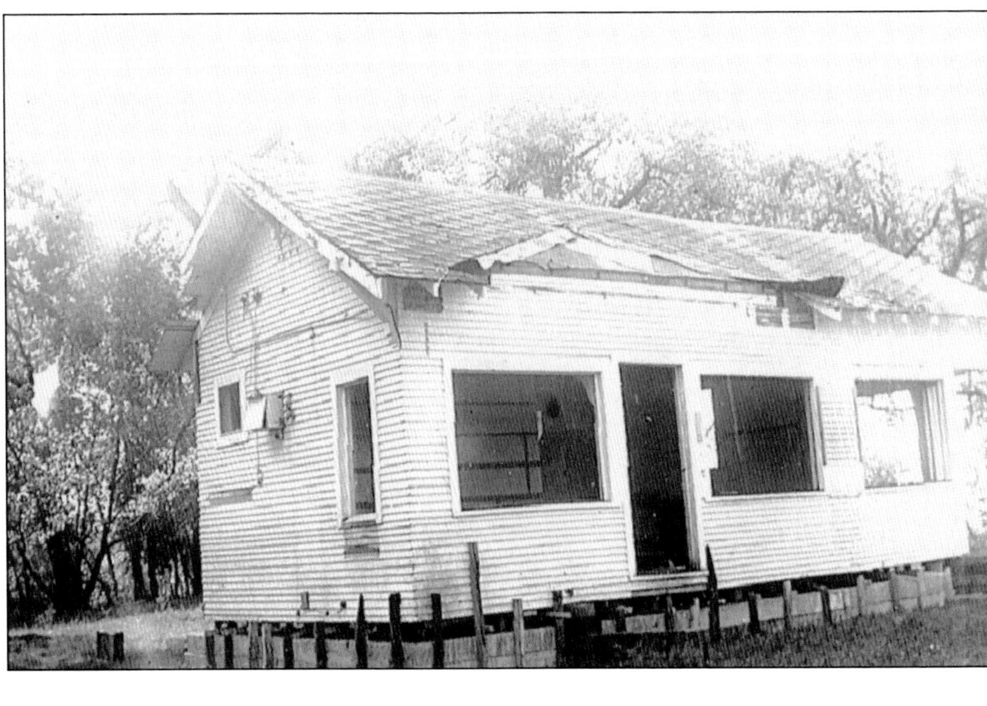

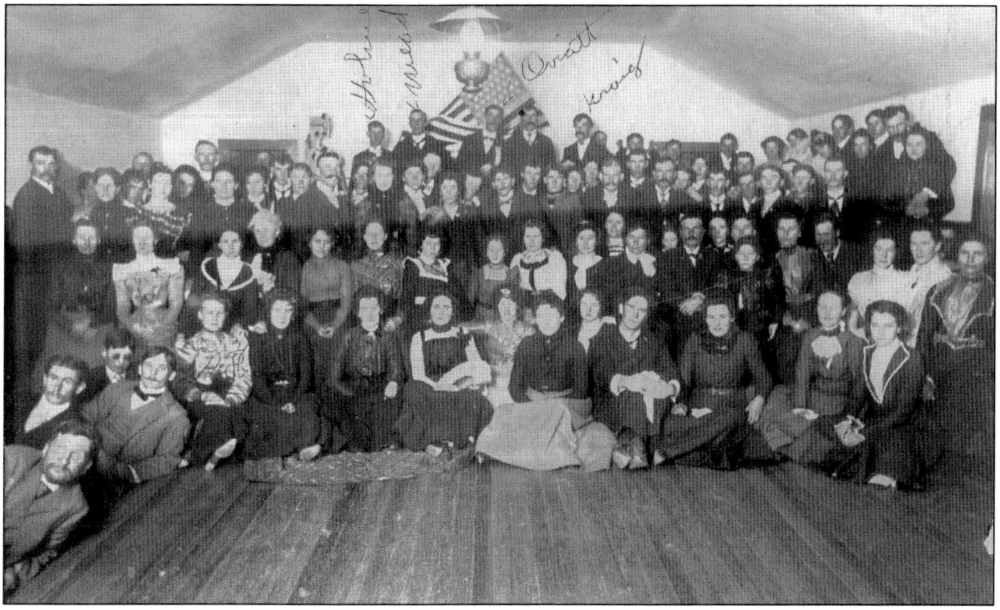

All was not just work; there was time for fun too. Sometime around 1900, a Farmer's Union Hall was built across the road from the M.S. Adams and Company Store. Saturday night dances were regularly held on the second floor of the building, and people came from as far away as Longmont to dance the night away. This photograph was taken around 1903. (Mary Mead Jensen family collection, Historic Highlandlake.)

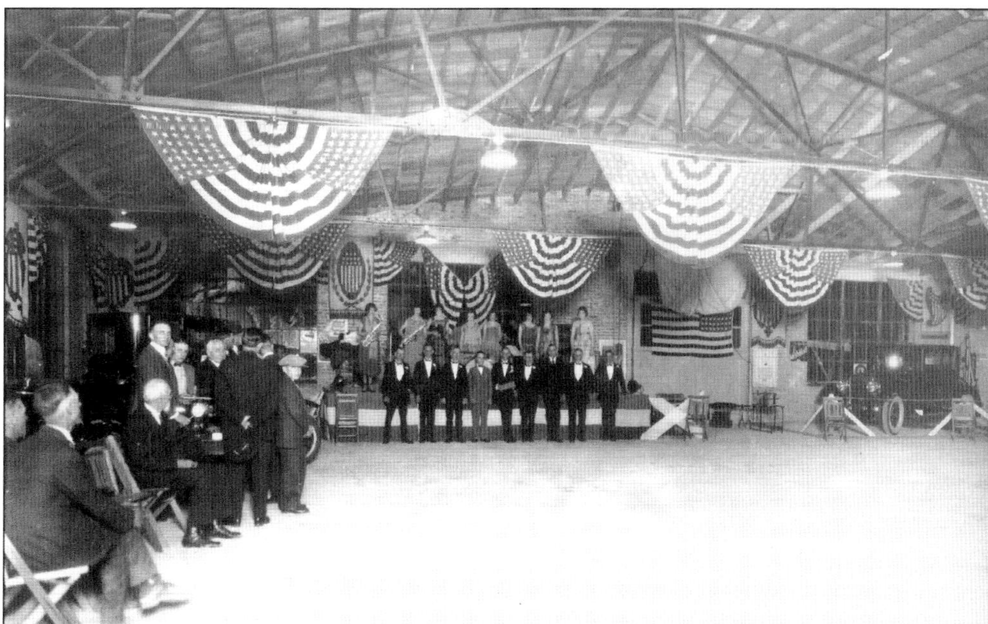

Quite a few of the early settlers in Highlandlake were Civil War veterans who later became members of a veterans' organization called the Grand Army of the Republic, or GAR. One of the highlights of the year included traveling to the annual encampments. The 1928 National Encampment was closer to home than most—just an easy drive or train ride to Denver. (Mary Mead Jensen family collection, Historic Highlandlake.)

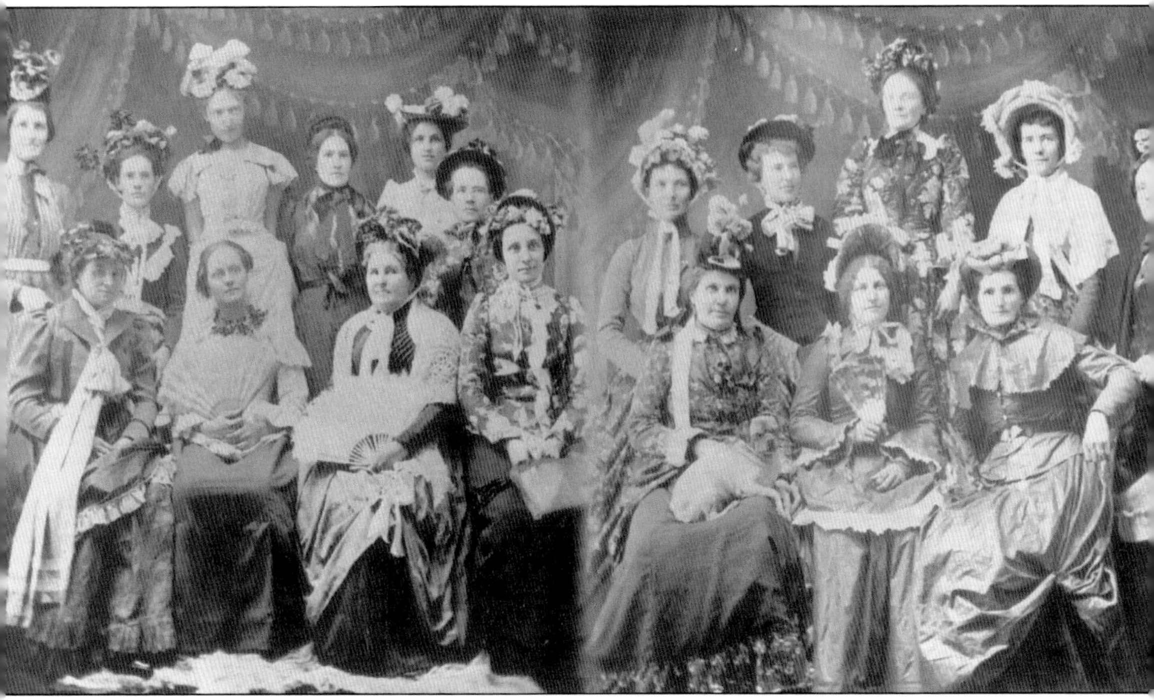

In 1897, several Highlandlake women formed the Women's Pleasure Club. They met monthly in member's homes where they had intellectually stimulating programs. In addition to educational programs, they visited museums, attended cultural events, and even went camping in the mountains. For this formal portrait, they dressed up in clothing from different time periods. The club became so popular that it soon expanded to other communities, including Liberty, Mount Zion, and Rinn. This photograph is actually two images merged. Pictured are, from left to right (front row) Mrs. Redd, Mrs. Dudrow, Mrs. Parks, Mrs. A. D. (Fanny D.) Holt, Mrs. Arnold, Mrs. John (Catherine) Reese, Mrs. William (Jenny) Mulligan, and Mrs. McMillan; (back row) Mrs. Carl A. (Ottilia) Carlson, Mrs. Parks, Mrs. Joe Bradley, Mrs. Arnold, Mrs. (Bertha) Morrison, Mrs. Sam (LaVina) Brinkerhuff, Mrs. Albert (Eva Maude) Peters, Mrs. Barr, Mrs. John (Maude) Thompson, and Mrs. John F. (Martha Ann) Bishop. (Longmont Museum and Cultural Center.)

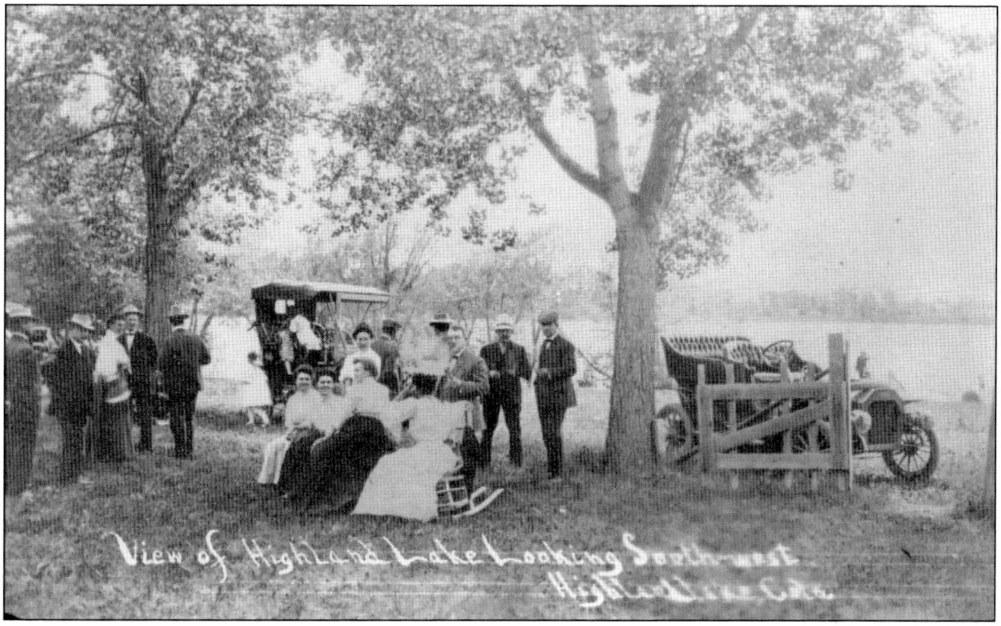

Almost from the beginning, Highlandlake took on a resort-like status. One could pitch a tent along the shore of the lake or, for a little more money, rent a small sleeping cabin or board at the Gateley's Fairview Farm Hotel. Visitors enjoyed picnicking, boating, fishing, tennis and croquet courts, and of course, the Saturday night dances. For city folks unfamiliar with country living, some residents offered tours of local farms, where visitors could learn to milk a cow, gather eggs, and pick berries. After the first car arrived in Highlandlake in 1903, even more visitors took advantage of Highlandlake's amenities. These picnickers brought their own rocking chairs. Below, in 1918, Highlandlaker Stella Doughty True takes her two sons, 11-year-old Clyde and 7-year-old Kenneth, for a boat ride on the lake. (Above, Historic Highlandlake; below, Mary True-Farr collection, Historic Highlandlake.)

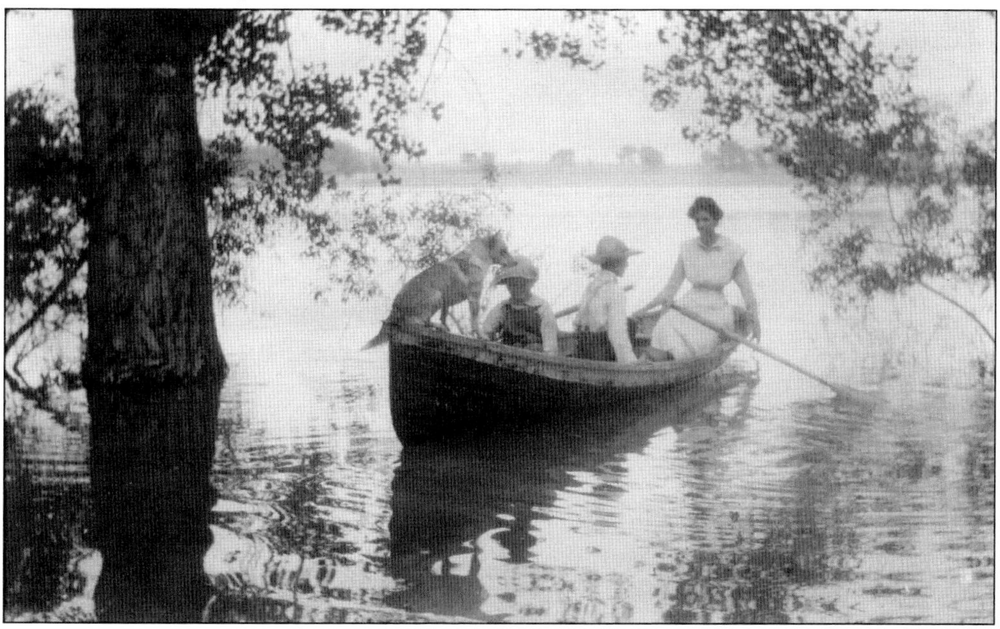

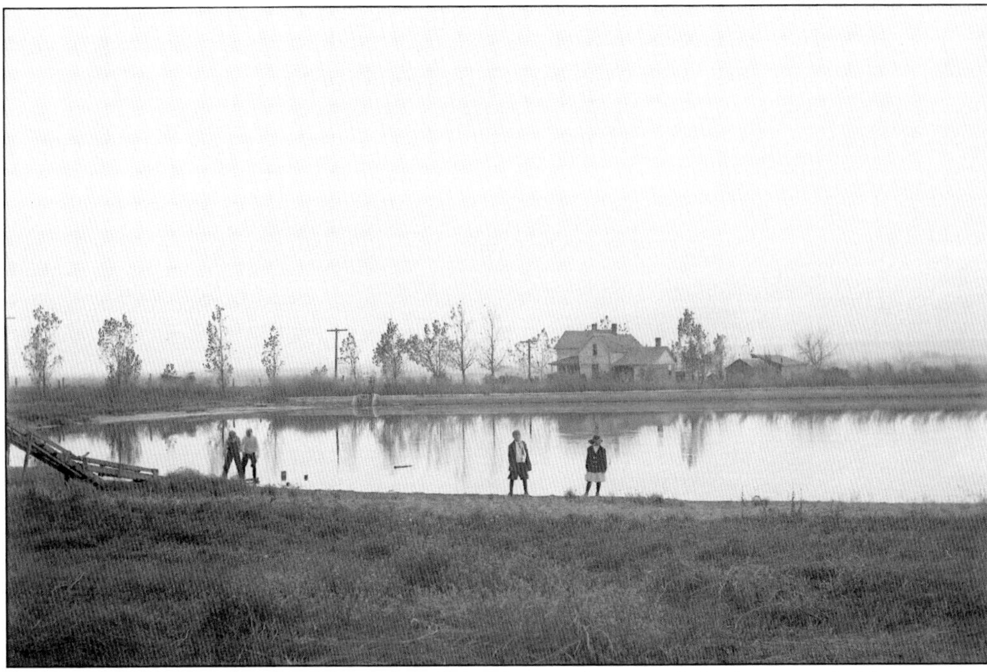

Four Highlandlake schoolchildren play along the southeastern lakeshore, probably sometime in the 1910s. William Oviatte, who was also the first postmaster, built the house around 1882. Residents used the ramp-like structure in the foreground in the winter to load large blocks of ice cut out of the lake onto wagons. (Longmont Museum and Cultural Center.)

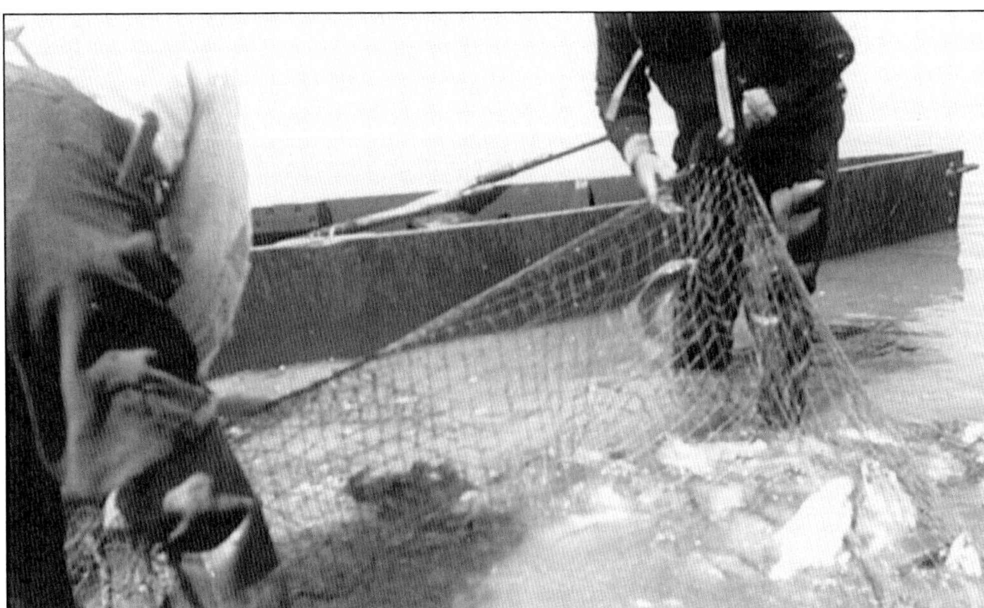

Fishermen drag a net through the lake to catch carp. The "good" fish are thrown back into the water, while the carp are fed to the hogs. Carp are fast growers and can quickly crowd out the more desirable fish, including trout, crappies, bass, blue gill, and perch. (Mary Mead Jensen family collection, Historic Highlandlake.)

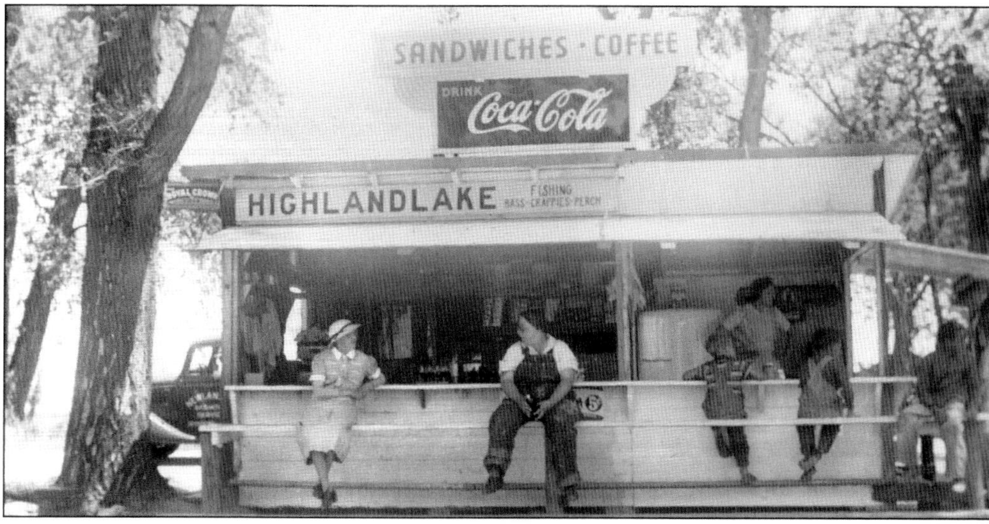

After a morning of fishing, nothing was better than heading over to the concession stand and ordering up a grilled hamburger and a little congenial conversation. Forget or run out of bait? One could buy that and, if needed, rent a boat too. Kids especially enjoyed the selection of candy and ice cream bars. (Mary Mead Jensen family collection, Historic Highlandlake.)

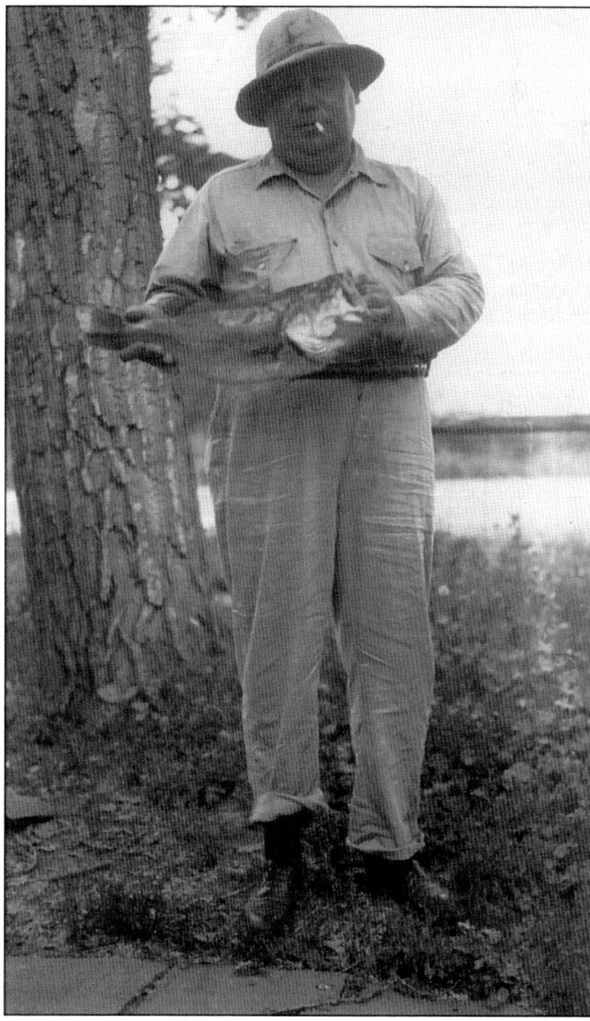

Almost from the time that Lorin C. Mead first homesteaded on the lakeshore, people recognized the lake as a prime fishing spot. In the 1920s, his son, Malcolm Mead, upgraded the lake's facilities by adding a café, concession stand, and boat rentals. In this photograph, dated about 1949, an angler proudly displays his largemouth bass. (Mary Mead Jensen family collection, Historic Highlandlake.)

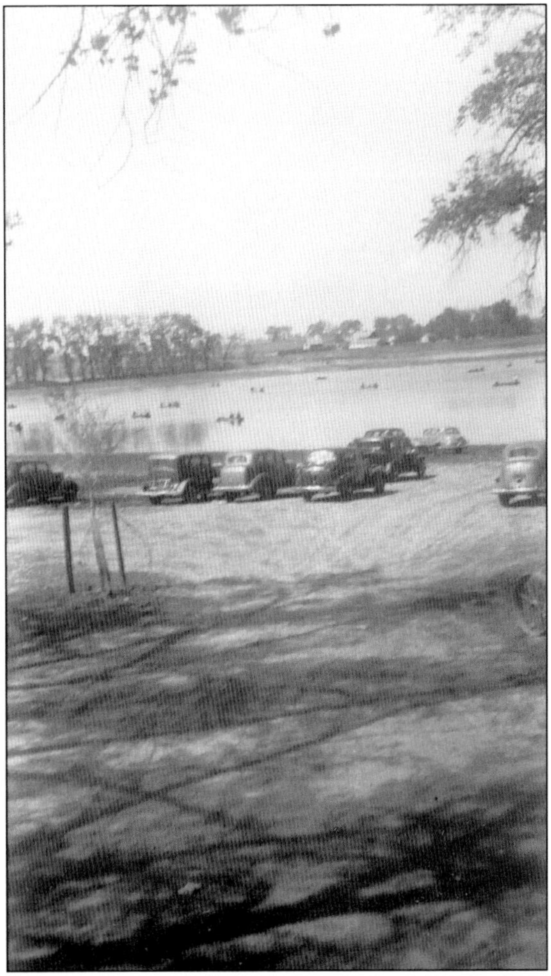

People came from all over to cast their lines into the fertile waters of Highland Lake. Malcolm Mead and, later, his daughter Mary Mead Jensen kept the lake well stocked with a variety of fish. On the day this c. 1942 photograph was snapped, there almost seem to be more boats than lake. (Mary Mead Jensen family collection, Historic Highlandlake.)

Tornados are rare in the Highlandlake-Mead region. However, in the summer of 1928, a twister ripped through the area, destroying parts of the Eckman farm. Pat Eckman, who was about 10 at the time, often remarked, "We didn't have much, but what we had, the tornado took away." This photograph shows some of the devastation. (Jean Eckman Nelson collection, Historic Highlandlake.)

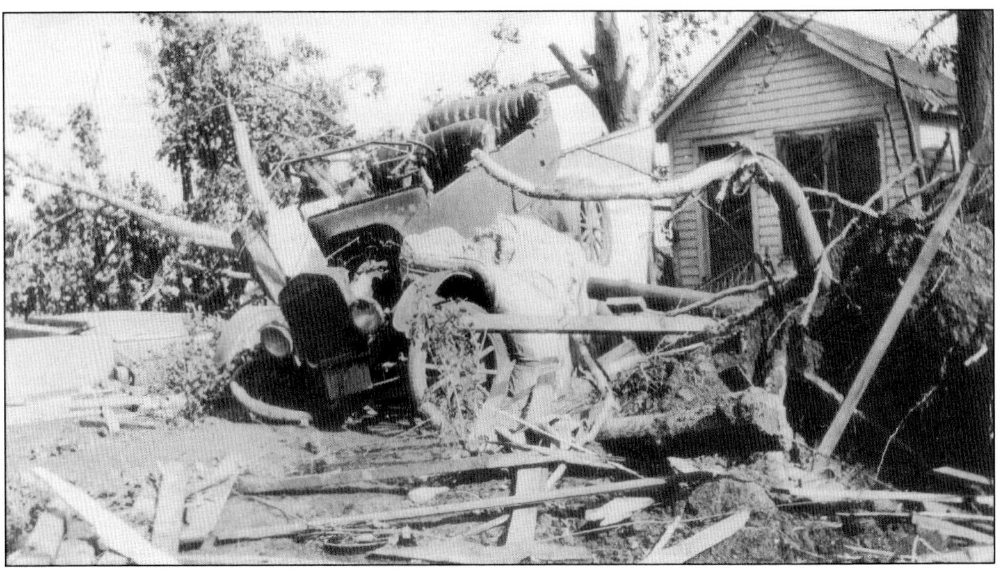

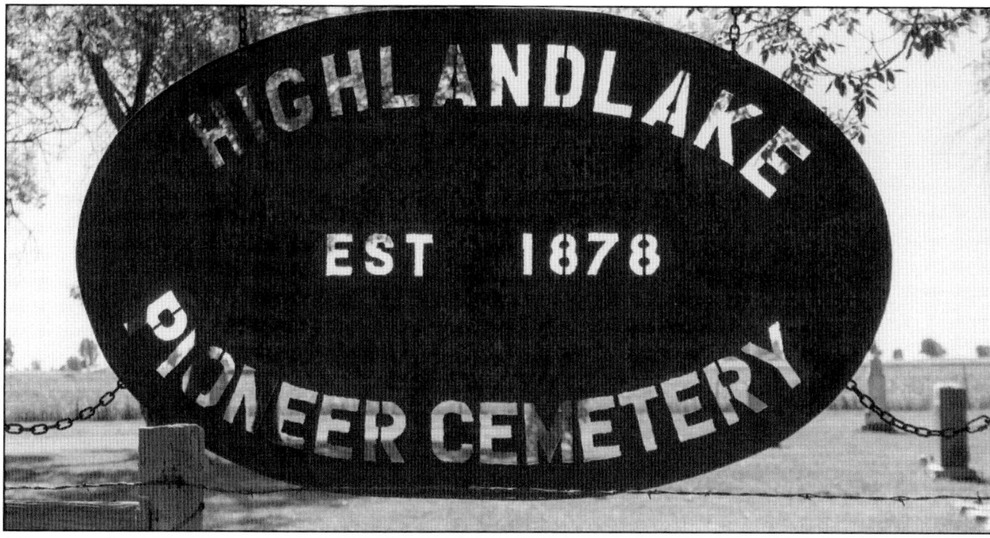

The metal sign at the entrance to the Highlandlake Pioneer Cemetery was handcrafted and donated to the cemetery by Highlandlake resident Ron Wood. Some of the fiercest fighting in the Pacific campaign of World War II was on the island of Iwo Jima, as described below. On February 28, 1945, the realities of that war came home, when word arrived that 19-year-old Marine private Richard "Dewey" Brossman was killed in action. The Highlandlake, Mead, and rural Longmont communities deeply mourned the loss of this likeable young man. Brossman is buried in the National Memorial Cemetery of the Pacific, Honolulu, Hawaii. (Above, Pauli Driver Smith collection, Historic Highlandlake; below, Jean Eckman Nelson collection, Historic Highlandlake.)

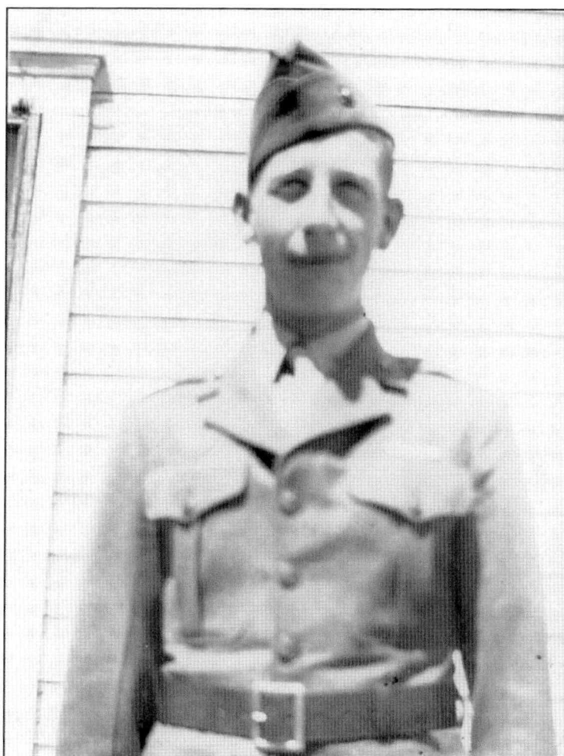

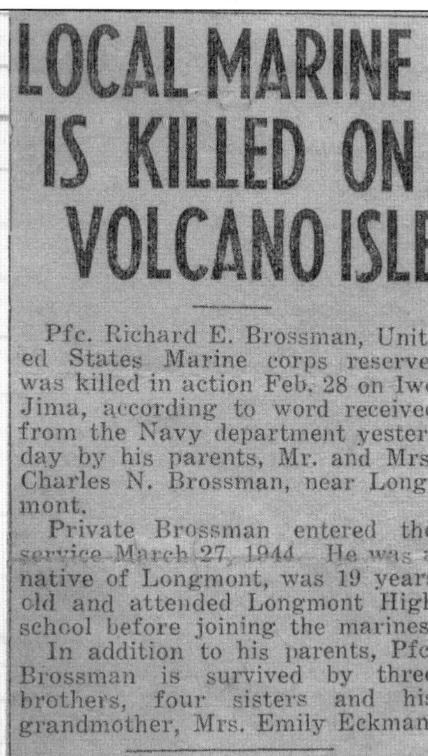

LOCAL MARINE IS KILLED ON VOLCANO ISLE

Pfc. Richard E. Brossman, United States Marine corps reserve, was killed in action Feb. 28 on Iwo Jima, according to word received from the Navy department yesterday by his parents, Mr. and Mrs. Charles N. Brossman, near Longmont.

Private Brossman entered the service March 27, 1944. He was a native of Longmont, was 19 years old and attended Longmont High school before joining the marines.

In addition to his parents, Pfc. Brossman is survived by three brothers, four sisters and his grandmother, Mrs. Emily Eckman.

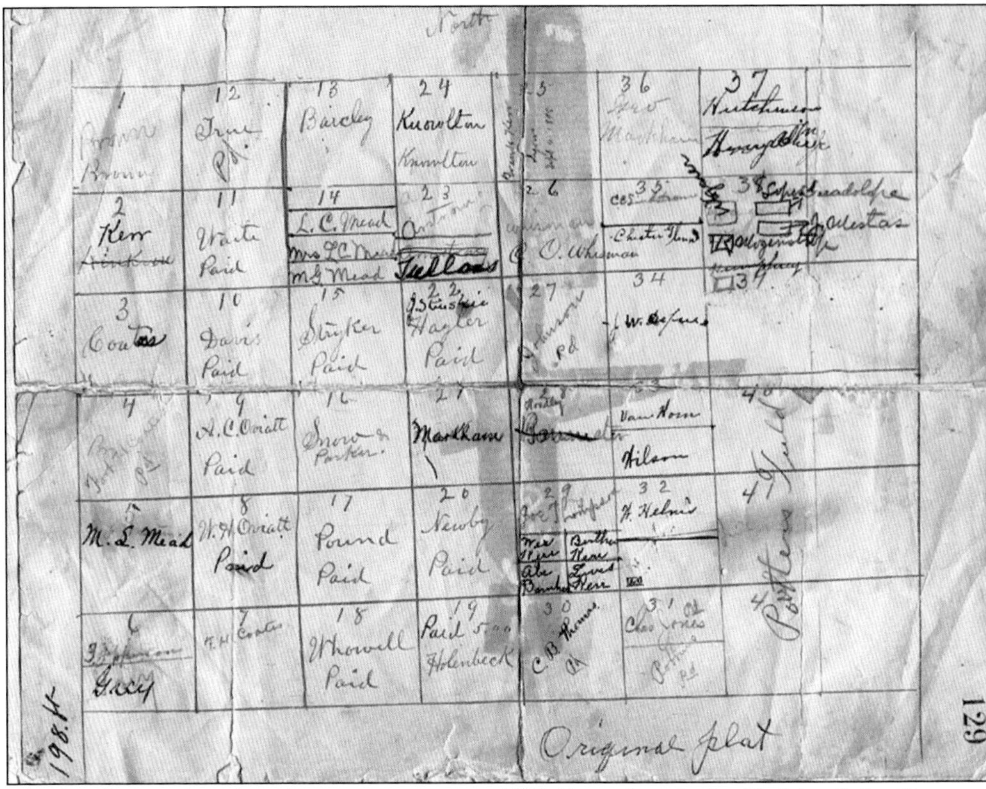

The first known burial in the cemetery was in 1878. Even though the Highlandlake Cemetery Association dissolved in 1906, descendants of Lorin C. Mead, one of the cemetery's original trustees, carefully preserved the records now located in the Historic Highlandlake, Inc., archives. Among the cemetery records is this hand-drawn plot map dating to 1883. Shown below is an early stock certificate to a cemetery lot issued to C.V. Stryker. One full lot had room for four to six graves. When Cornelius V. Stryker purchased this lot in 1884, it already held the grave of his son Alvin, who died at age 15 on December 10, 1880. Both Cornelius and his wife, Ursula, are buried next to their son. The stock certificate was donated by Mark French to the Historic Highlandlake Archives. (Both, Historic Highlandlake.)

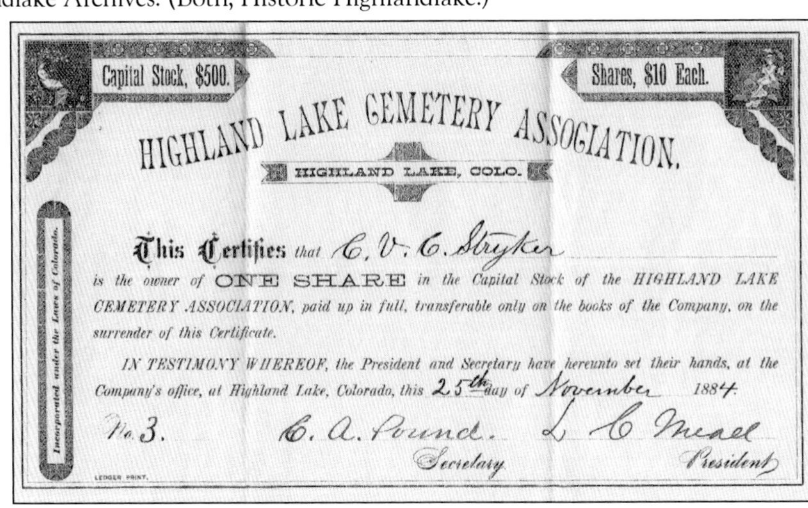
Capital Stock, $500.

Shares, $10 Each.

HIGHLAND LAKE CEMETERY ASSOCIATION.

HIGHLAND LAKE, COLO.

Incorporated under the Laws of Colorado.

This Certifies that C. V. C. Stryker is the owner of ONE SHARE in the Capital Stock of the HIGHLAND LAKE CEMETERY ASSOCIATION, paid up in full, transferable only on the books of the Company, on the surrender of this Certificate.

IN TESTIMONY WHEREOF, the President and Secretary have hereunto set their hands, at the Company's office, at Highland Lake, Colorado, this 25th day of November 1884.

No. 3.

C. A. Pound. Secretary

L. C. Mead President

LEDGER PRINT.

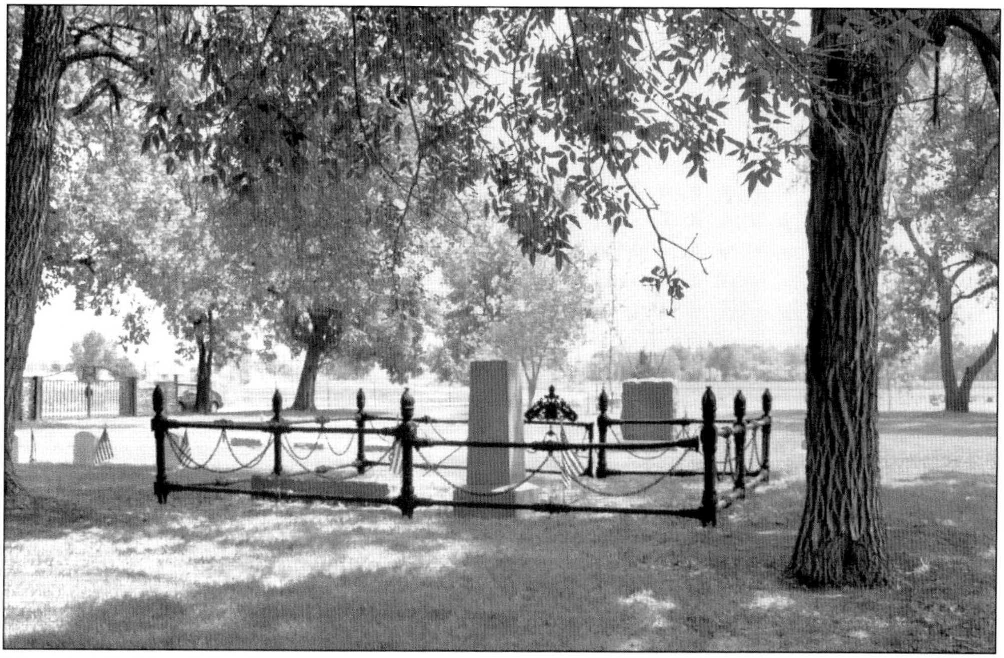

Historic Highlandlake, Inc., the historical society for the Highlandlake and Mead area, now owns and maintains the Highlandlake Pioneer Cemetery. In 2000, Lawrence Jensen, the great-grandson of Lorin and Elizabeth Mead, generously donated the cemetery deed to the society. This scene looks over the George Davis lot, the only one in the cemetery that has an iron fence around it. Right, whenever the name of someone in an unmarked grave is identified, volunteer Mark Parsons makes a headstone for him or her. As seen here, Parsons is kneeling behind the white marble headstone he made for the Hagler baby who died in 1888. (Both, Pauli Driver Smith collection, Historic Highlandlake.)

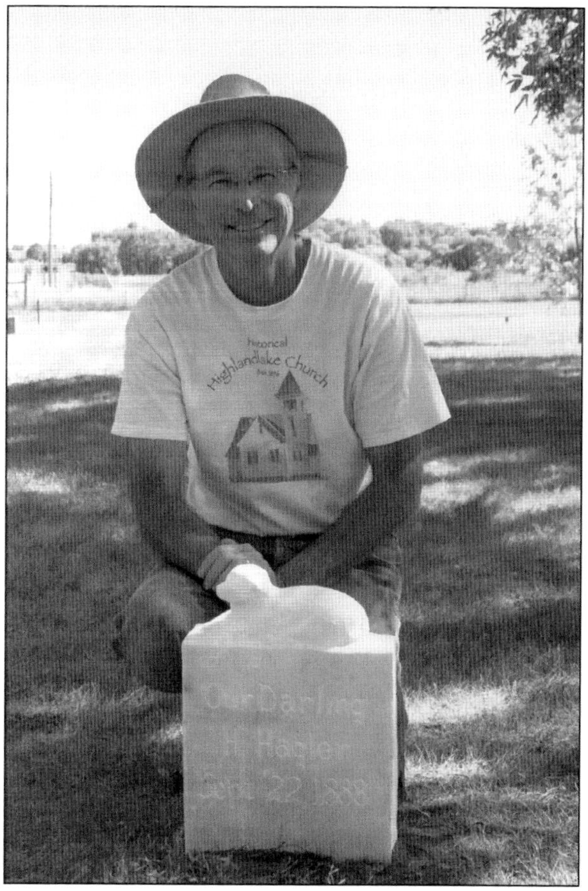

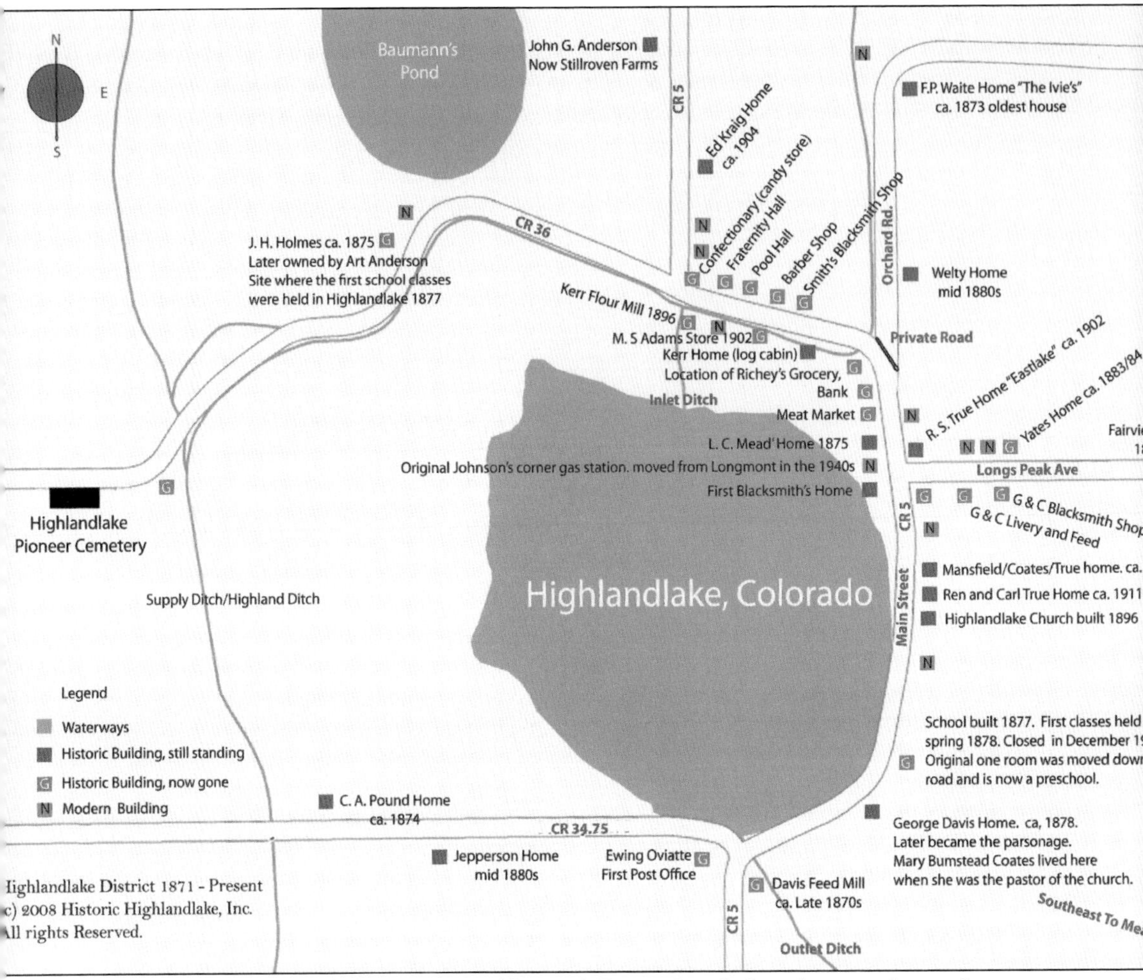

This map of Highlandlake was created by Pauli Driver Smith from historical records, newspaper clippings, and personal recollections of longtime Highlandlake families. (Pauli Driver Smith collection, Historic Highlandlake.)

Two

The Little Town that Could

The town of Mead has its roots in the last-minute decision of the Great Western Railroad to reroute its planned tracks, highline beet dump, and train station from Highlandlake to the eastern border of Paul Mead's (nephew of Lorin C. Mead) farm, almost two miles away. Since the new beet dump was on the border of his land, Mead, at the advice of some business associates, along with his wife, Ariet Palmer Mead, laid out several blocks and streets including space for a park and school. He started offering building lots for sale in December 1905. In the first week alone, he sold 35.

Mead's daughter Myra Imogene Mead Cope later reminisced, "I can remember the founding of the little town of Mead, Colorado. In 1906, Mother and Father laid out the plans for the town and named the streets. They wanted it to be a good town, with no 'evil element,' and had it stipulated in the titles to all the lots that liquor was not to be sold on any of the property, ever."

At first, most stores and businesses were located on the east side of Old Main (Third) Street since that side of the road was not subject to the dry clause in the property deeds. In 1912, Main Street moved to Fourth Street. At its peak, Mead had a library, two doctors' offices, three general stores, a hotel, a combination grocery and meat market, two saloons, butcher shop, drugstore, post office, lunchroom, pool hall, two banks, a newspaper, and bowling alley. Additionally, there were three filling stations, two auto garages, a farm implement dealer, two livery stables, a lumberyard, three blacksmiths, a harness repair shop, newspaper, Ford car dealership, pickle factory, hay mill, and pea hulling factory.

The Great Depression of the 1930s dealt a crushing blow to the town. For almost 60 years, the town struggled to overcome the Depression's debilitating effects until in the 1990s, when families once again discovered Mead's quaint charm and rural beauty. Today, Mead is a growing, vibrant "Little Town with a Big Future."

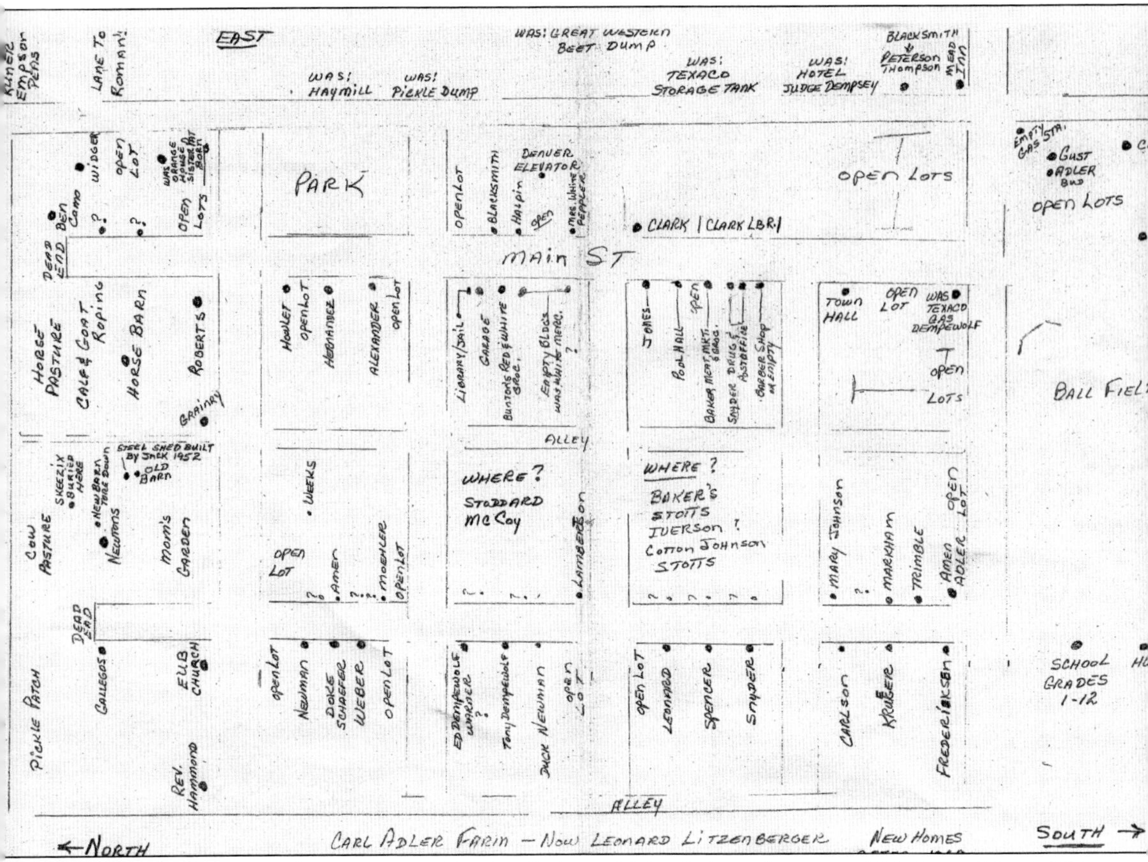

Gerald Newton, who grew up in Highlandlake and Mead, drew this map of 1940s-era Mead. This remarkably detailed record shows both the locations of businesses and where many Mead families were living during this period. (Dale and Patricia French collection, Historic Highlandlake.)

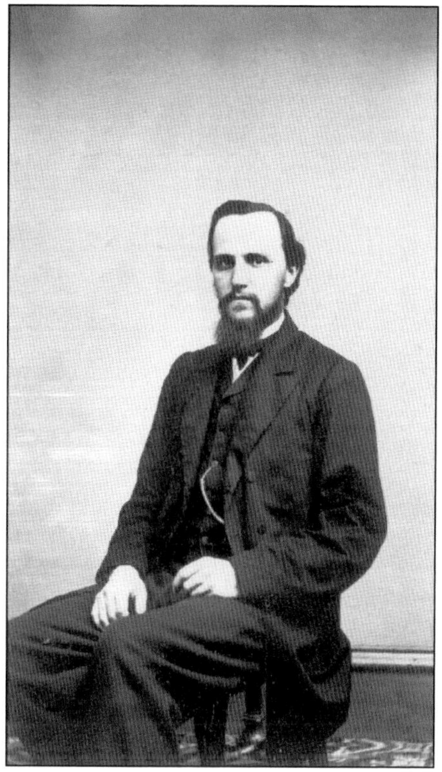

Dr. Martin Luther Mead, MD, shown here about the time he graduated from medical school in Albany, New York, came to Highlandlake in 1883 at the urging of his brother Lorin C. Mead. He built his home on a rise southeast of the lake on land that later became the town of Mead. Shown below is Dr. Mead as he appeared in his later years. Dr. Mead died on September 5, 1899, after suffering for over eight years from tuberculosis of the brain. He is buried in the Highlandlake Pioneer Cemetery next to his wife and three sons. (Right, Ralph Kirkpatrick; below, Mary Mead Jensen family collections, Historic Highlandlake.)

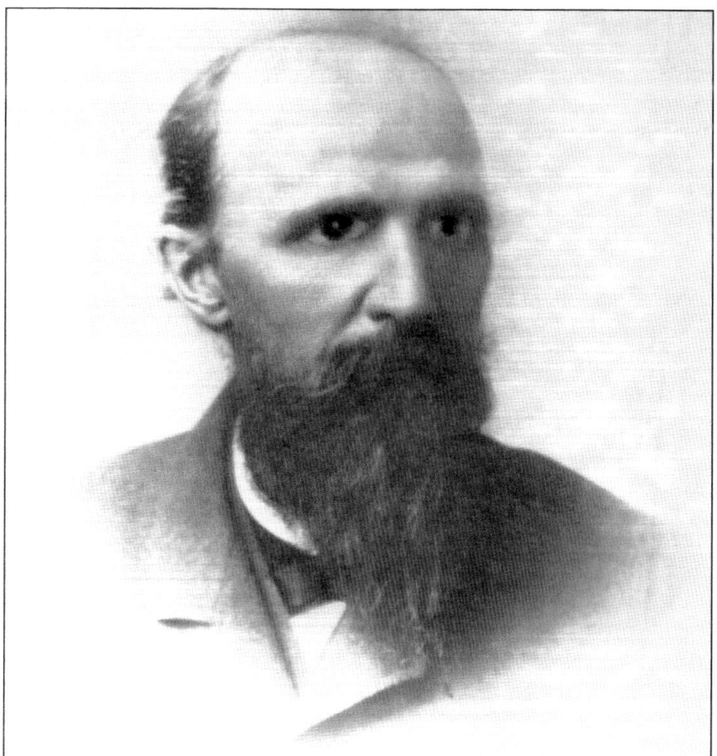

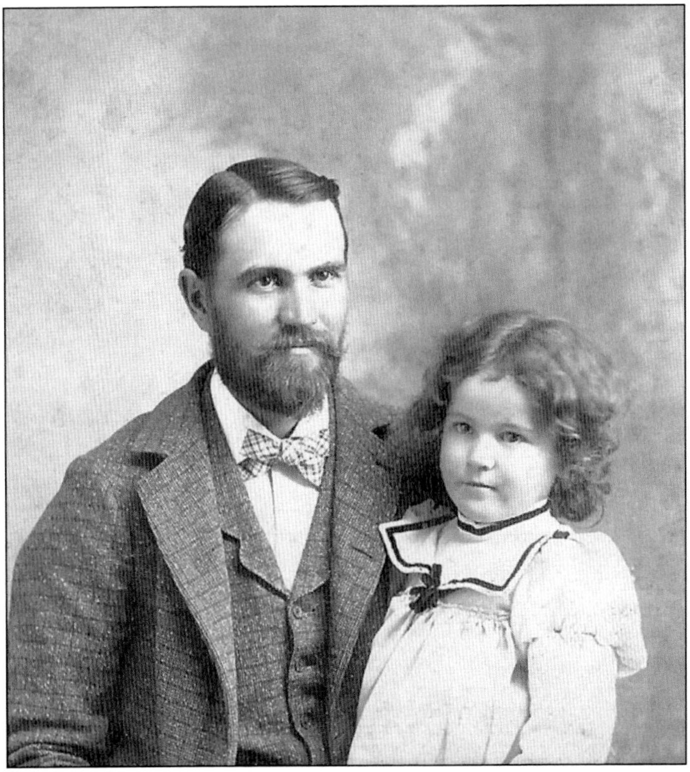

Paul Martin Mead, the founder of the town of Mead, came to Highlandlake in 1883 with his parents and two younger brothers, Harry and Fred. His father, Dr. Martin Luther Mead, was the first doctor in the area and the brother of Highlandlake founder Lorin C. Mead. Paul is shown here with his eldest daughter Myra Imogene Mead Cope. (Ralph Kirkpatrick collection, Historic Highlandlake.)

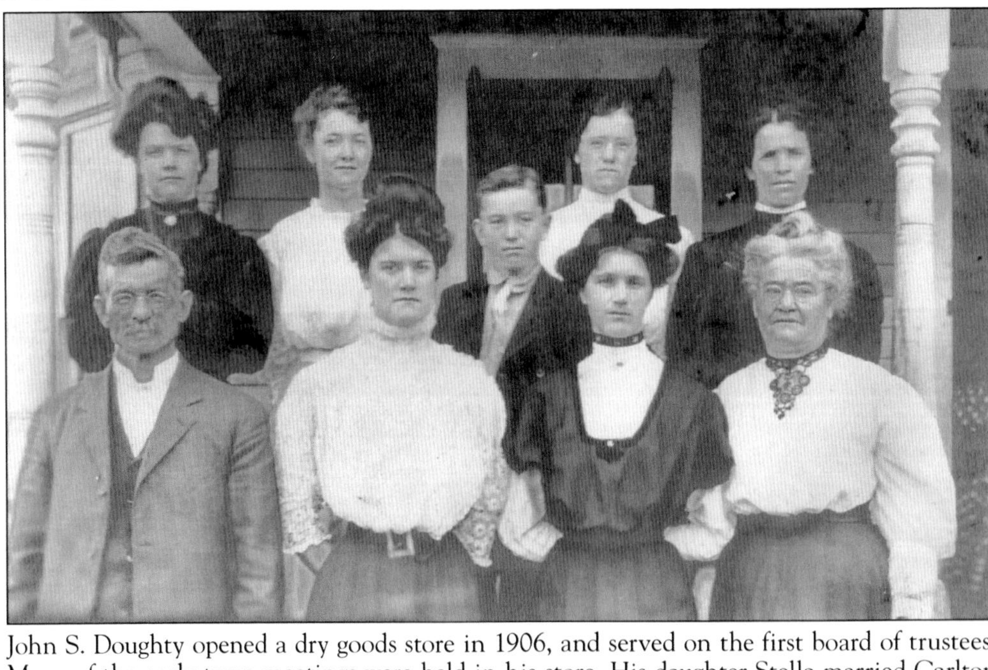

John S. Doughty opened a dry goods store in 1906, and served on the first board of trustees. Many of the early town meetings were held in his store. His daughter Stella married Carlton True. Those pictured are, from left to right, (first row) John, Ada, Stella, and John's wife, Helen; (second row) Edna, Mattie, Samuel "Sammie," Bertha, and Maybel. (Mary True-Farr collection, Historic Highlandlake.)

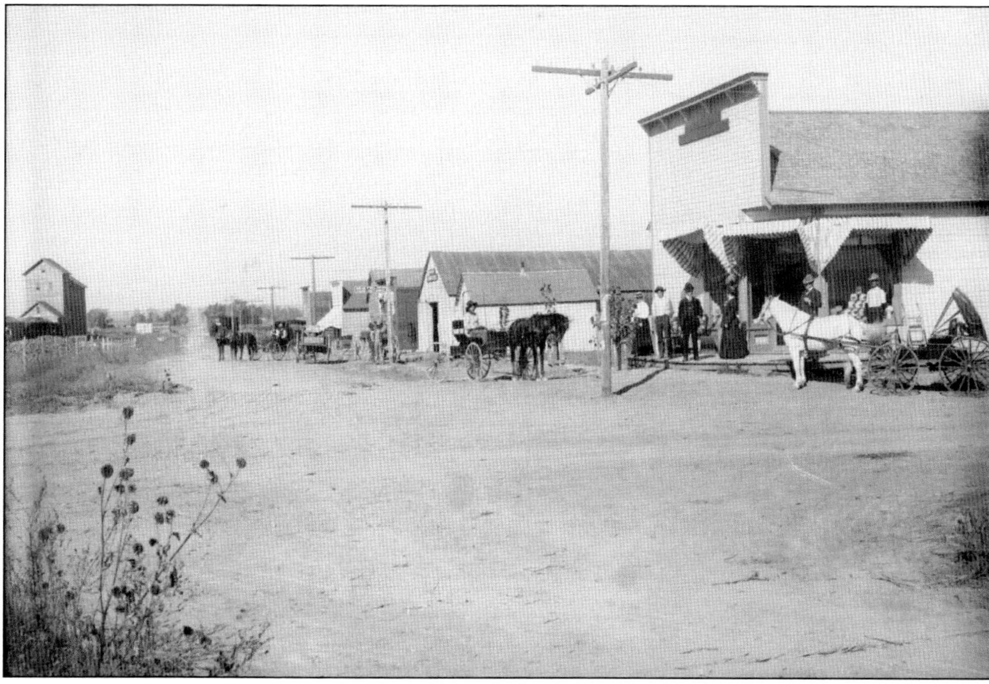

This is one of the earliest known photographs of the town of Mead. It was taken around 1908 in the intersection of what is now Welker Avenue and Weld County Road 7. Marion S. Adams (wearing the dark suit and black hat) is standing in front of his store next to his wife, Rebecca. Marion was Mead's first mayor. (Longmont Museum and Cultural Center.)

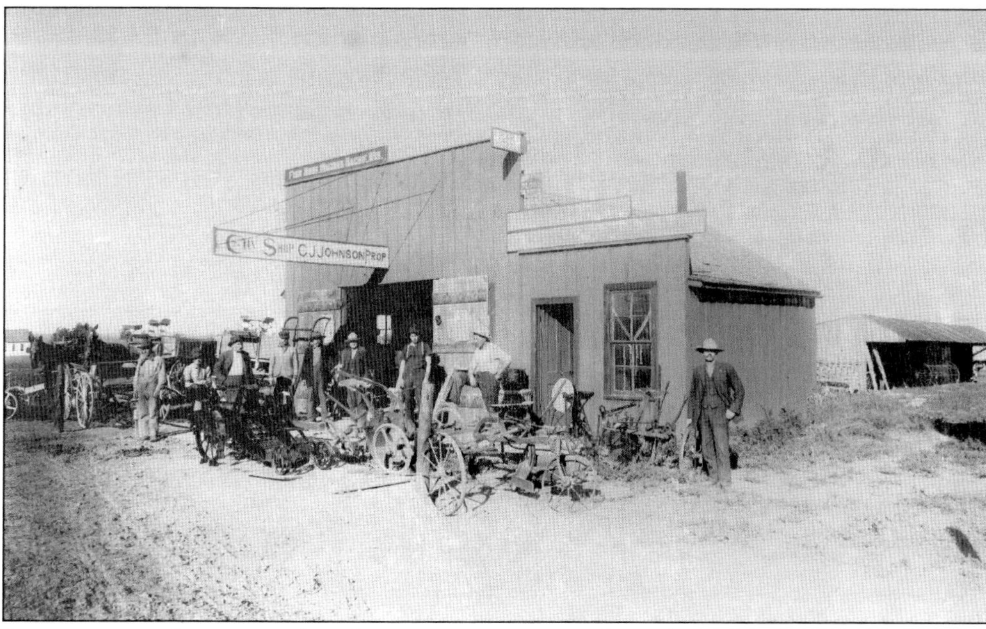

As soon as word went out that there was a new town along the railroad tracks southeast of Highlandlake, Charley Johnson started planning his move to Mead. Opening Johnson's City Blacksmith in 1906, Charley repaired and sold farm equipment and carried miscellaneous farming supplies. (MaryAnn Mulligan collection, Historic Highlandlake.)

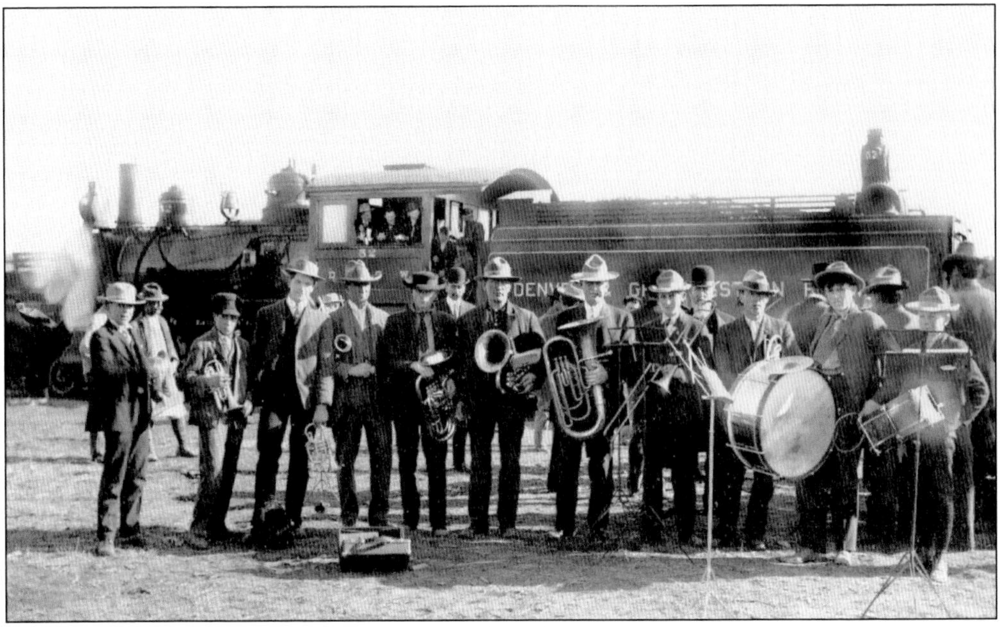

On January 18, 1907, about 90 businessmen boarded the Denver & Great Western train in Longmont for a "Boosters Excursion." The daylong trip took them through what the *Longmont Ledger* later described as "the best sugar beet country in the world." When they arrived at their first of three stops, the Mead Band, along with several prominent citizens, met and escorted them through town. (Longmont Museum and Cultural Center.)

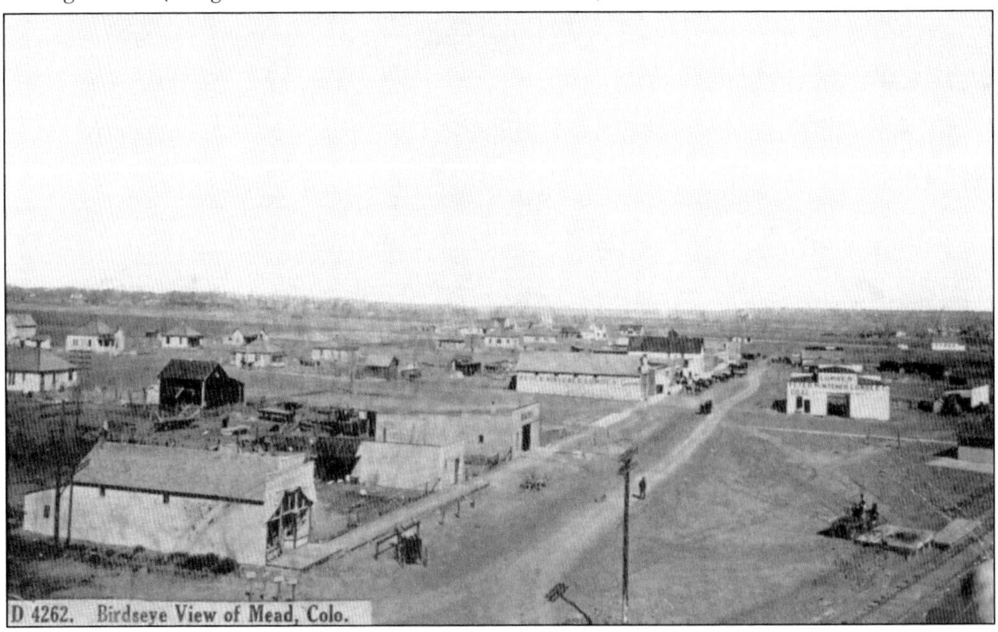

The Denver Elevator Company owned three elevators in Mead. The southernmost one was located where the post office is now and was a popular vantage point for taking photographs of the town. Peter Mintener Lumber is on the right. The building directly across, with Peter Mintener's name on it, sold in 1911 and became the Rose Hall, where community dances and meetings were held. (Mark French collection, Historic Highlandlake.)

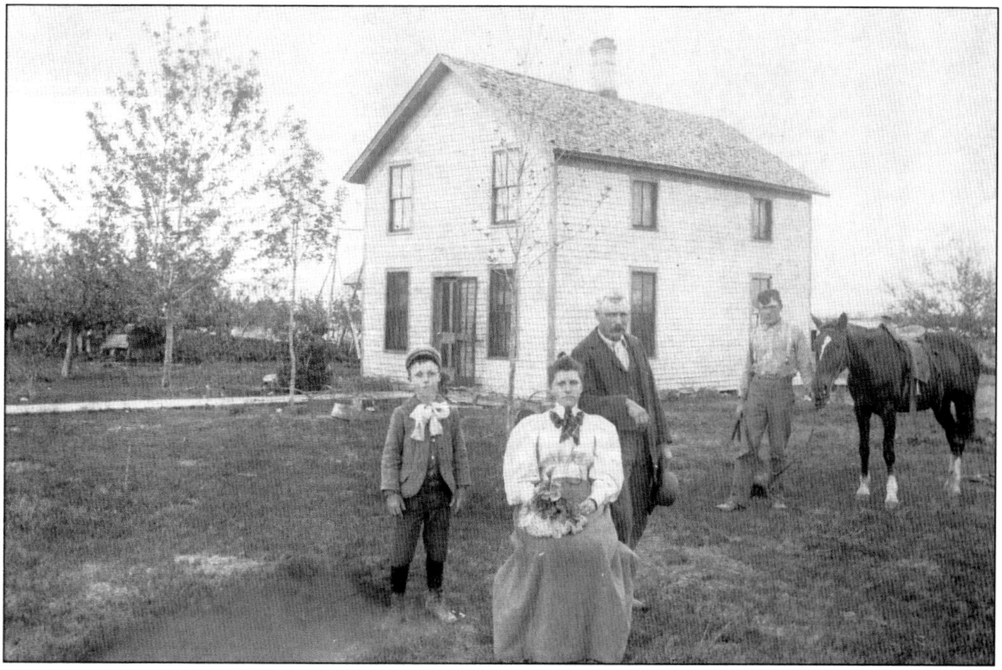

The Mulligan family homesteaded about 90 acres of land in 1874, making them one the earliest families in the Highlandlake-Mead area. A subdivision, appropriately known as Mulligan Estates, is now on part of their former farm. Devout Catholics, they were active members of the Guardian Angels Church in Mead. The photograph above was taken around 1894 and shows the second generation of Mulligans posing in front of their new home. From left to right are Mark Lewis Mulligan, Jennie Wilson Mulligan, William Mulligan, and Jennie's brother John Commodore Perry Wilson. Below, in the mid-1920s, the Mulligan and Turner girls cool off in Mulligan Lake. Veva Turner is third from the left. (Above, Mary Ann Mulligan collection, Historic Highlandlake; below, Alice York collection, Historic Highlandlake.)

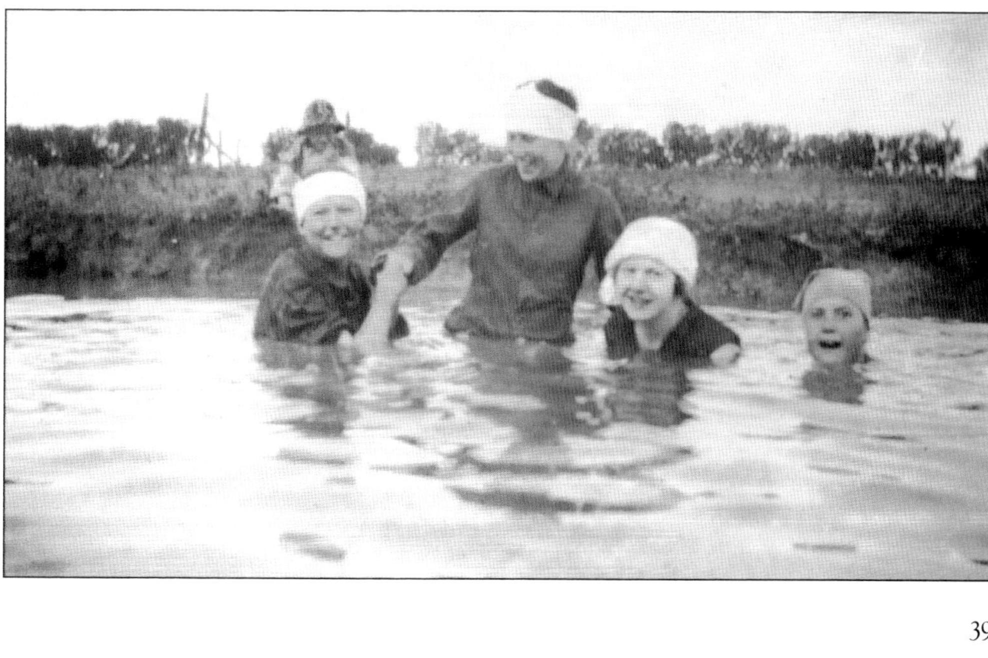

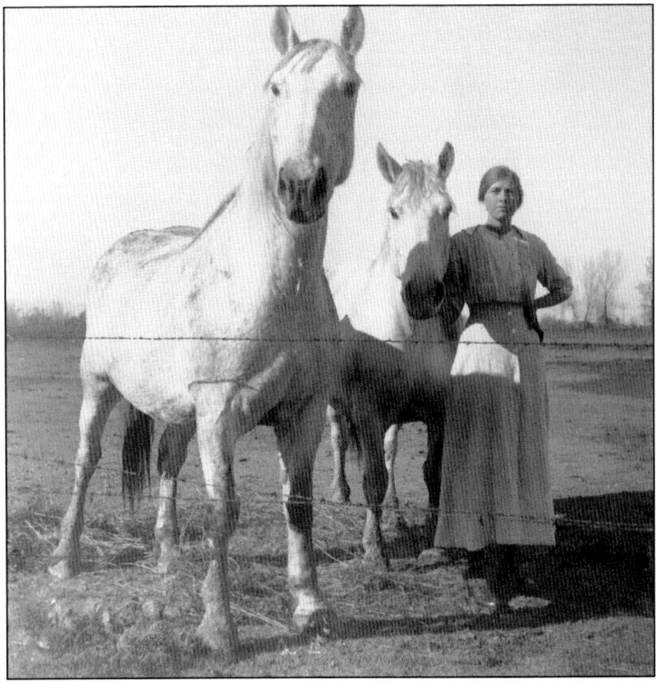

The families in the outlying farms surrounding Mead were important to the new town's economy. In this c. 1916 photograph, a young woman shows off her grey mare and yearling. One can see Mead in the background. Note the newly installed phone lines that connected the residents to each other and the outside world. (Longmont Museum and Cultural Center.)

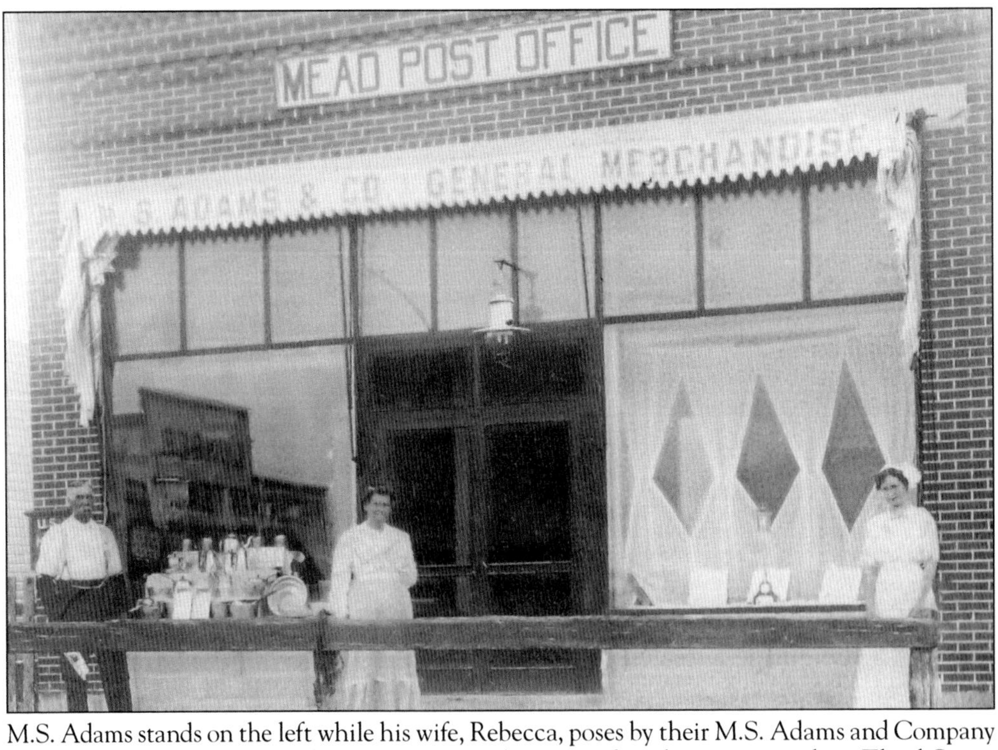

M.S. Adams stands on the left while his wife, Rebecca, poses by their M.S. Adams and Company Store's doorway. They built this store on Fourth Street after they outgrew their Third Street building. One can see Peter Mintener's Lumberyard reflected in the window. The M.S. Adams and Company Building is located on Fourth Street and is now a private residence. The woman ar right is unidentified. (Linda and Neil Gehring collection, Historic Highlandlake.)

Both Will Markham and his bride, Pearl Newby, were children of early Highlandlake pioneers. Like their parents, they were heavily involved in local affairs. Active in church and the schools, they first served for years on the Highlandlake School Board and then later the Mead Consolidated Schools Board. (Jean Howe collection, Historic Highlandlake.)

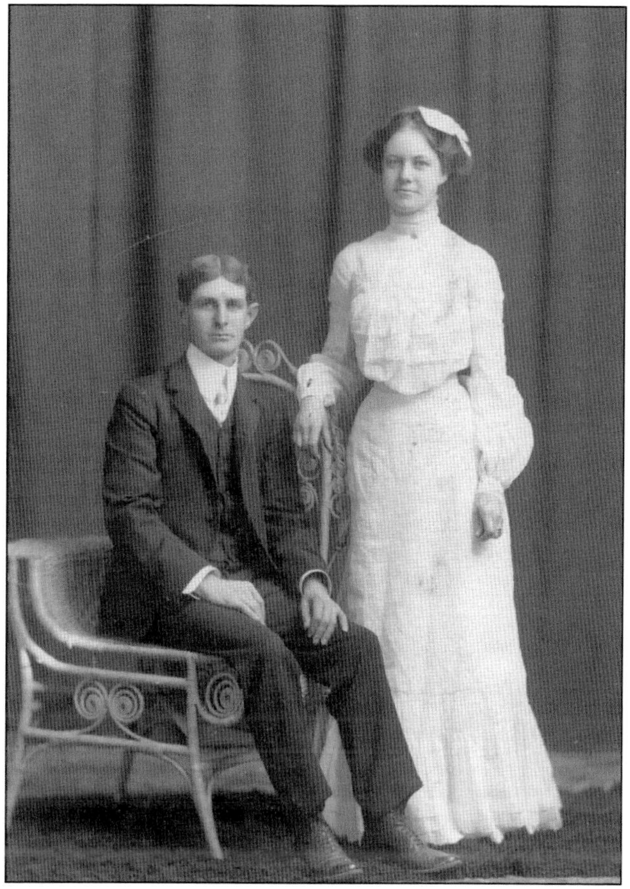

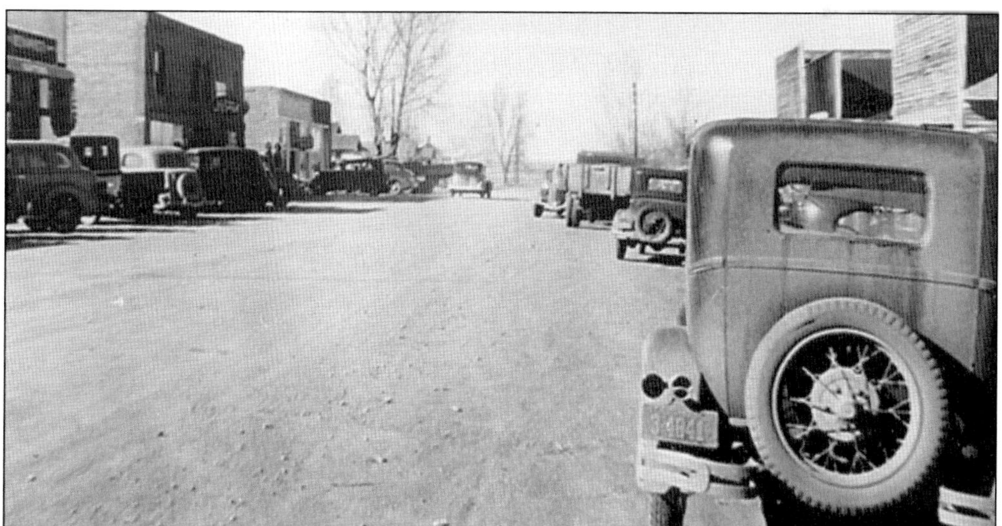

During the spring of 1912, substantial brick stores started going up on the west side of Fourth Street. The first on the left is McCormick's General Merchandise. Next is the Ford dealership. The town hall and library are farther down. The Farmer's and Merchant's Bank was once located in this block but went out of business in 1928. (Hester Markham collection, Historic Highlandlake.)

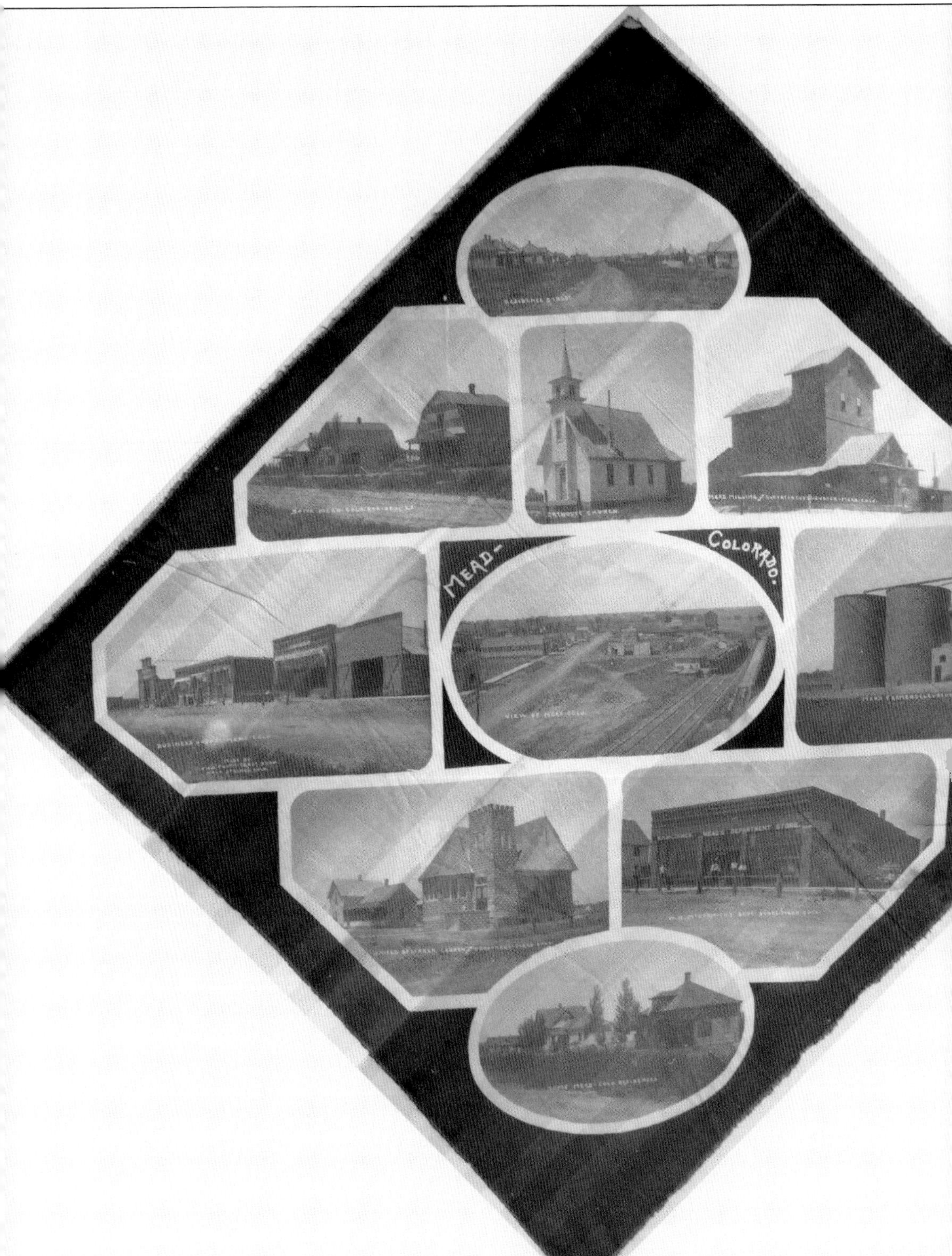
MEAD-
COLORADO.

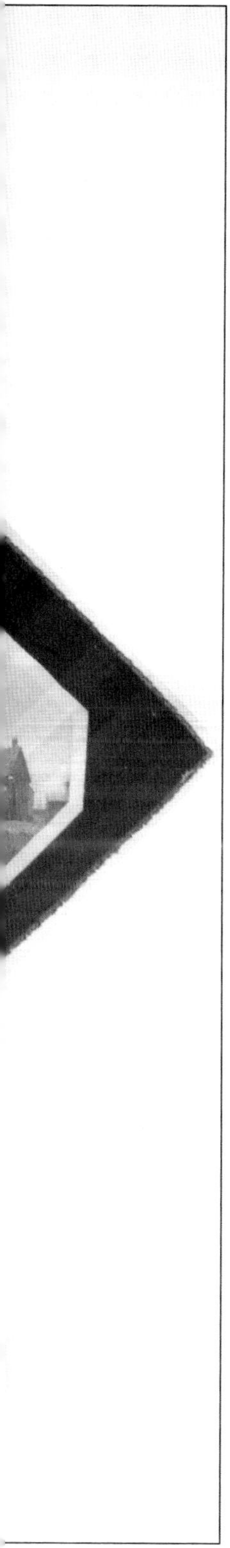

Sometime in the late 1910s or early 1920s, the Mead W.H. McCormick store ordered a shipment of custom-printed, cotton scarves from the Photo-Craft Shop in Colorado Springs. The oversized, cyanotype-printed souvenir head coverings featured 10 different photographs of stores and homes in Mead, all taken before 1915. None of the original photographs are known to exist today. Jerry Fruth currently owns this scarf, and it is on display in the Mead Town Hall. From left to right, the captions read, starting at top and going from top row down, as follows: "Residential Street," "Some Mead, Colo. Residences," "Catholic Church," "Mead Milling and Elevator Co's elevator," "Mead, Colo.," "Business–Houses, Mead, Colo.," "View of Mead, Colo.," "Mead Farmer's Elevator," "United Brethren Church and Parsonage, Mead, Colo.," "W.H. McCormick's Department Store. Mead, Colo.," and "Some Mead, Colo. Residences." (Jerry Fruth collection, Historic Highlandlake.)

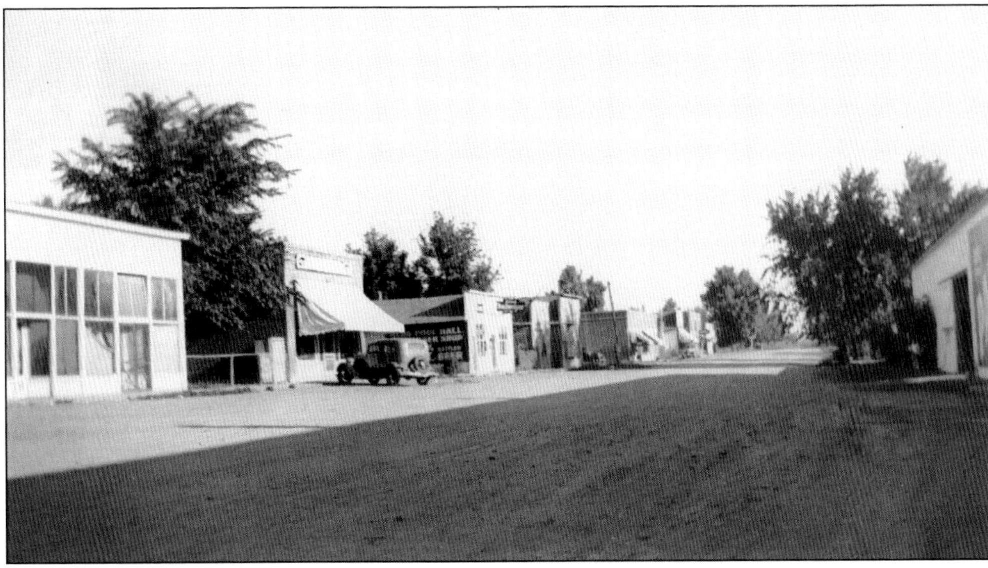

In this 1939 view of the south end of Main Street, Snider's Drugstore is on the left. Just beyond in the old M.S. Adams and Company Building is Baker's Grocery. Next is the combination pool hall and barbershop, and the last building on the left was a café. Clark's (formally, Peter Mintener) Lumber is on the right. (Hester Markham collection, Historic Highlandlake.)

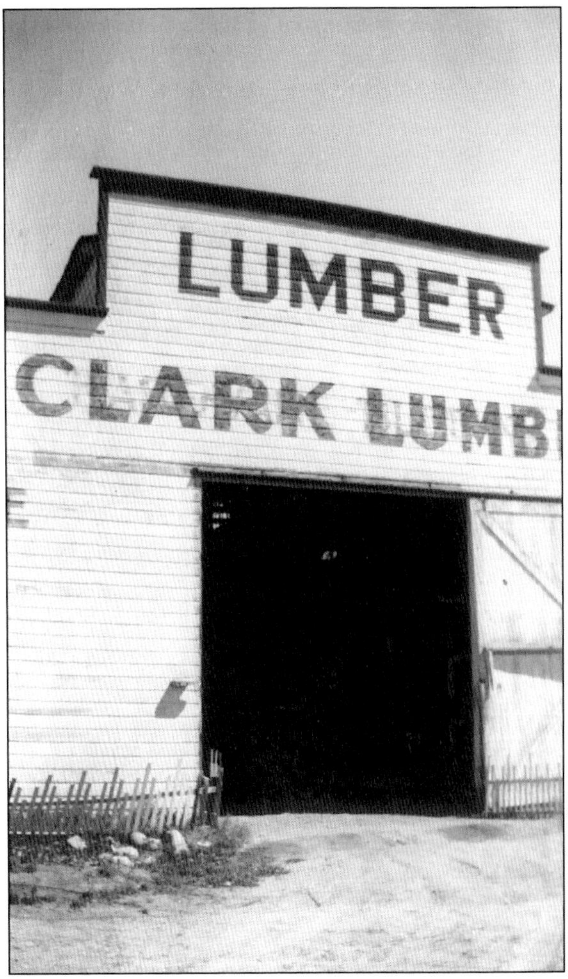

The Peter Mintener Lumberyard eventually sold out to the Clark family, who continued on the lumber business. During the summer, Saturday nights frequently featured movies projected onto the white, outside west wall of this building. Seating was "bring your own chair or blanket." The alternative was to sit on the sidewalk or street. (Robert Clark collection, Historic Highlandlake.)

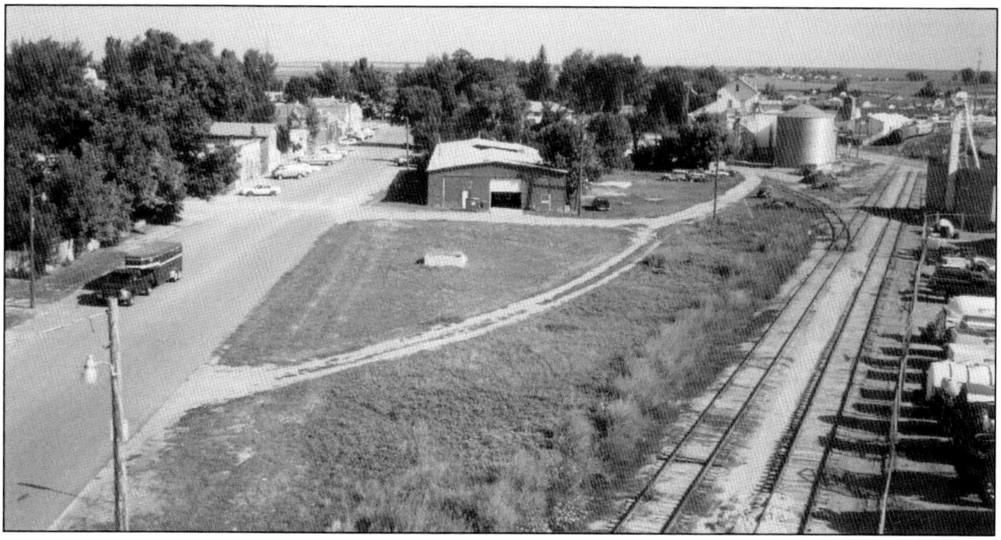

In 1999, several members of the Mead Area Community Foundation decided it would be fun to do a "then and now" photograph shoot. A board member working for United Power brought out a bucket truck and lifted Karen Rademacher up to about the height from which the original photographs were taken. As one can see in comparing this image to the one at the bottom of page 38, little had changed over the previous 100 years. (Karen Rademacher collection, Historic Highlandlake.)

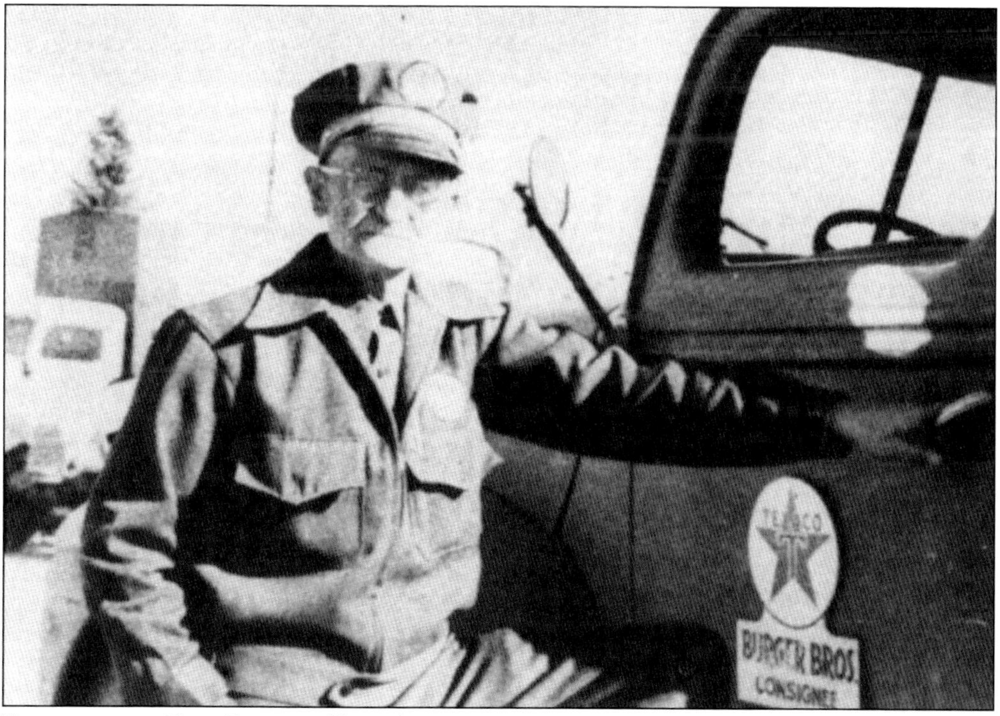

For many years, Tony Dempewolf ran the Texaco gas station, located near the southwest corner of Welker Avenue and Main Street (Fourth Street). In addition to running a full-service gas station, he also delivered gasoline to area farms. Dempewolf died on January 14, 1957, just a few years after this c. 1954 photograph was taken. (Mead Bulldog Annual collection, Historic Highlandlake.)

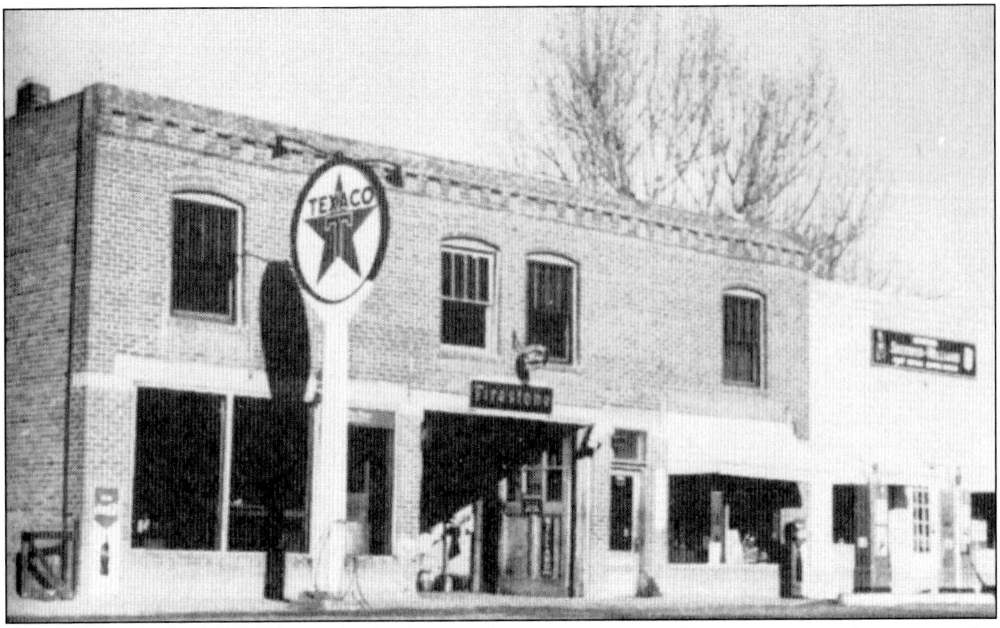

The 1950s-era Ford dealership, which also sold Firestone tires and Texaco gas, was located in the 300 block of Main Street. The town hall was also located in this block after it moved from the old assembly hall. The town hall and the volunteer fire department shared the same building, which once housed the Farmer's and Merchant's Bank. Across the road was Frank Boro's blacksmith shop. (Mead Bulldog Annual collection, Historic Highlandlake.)

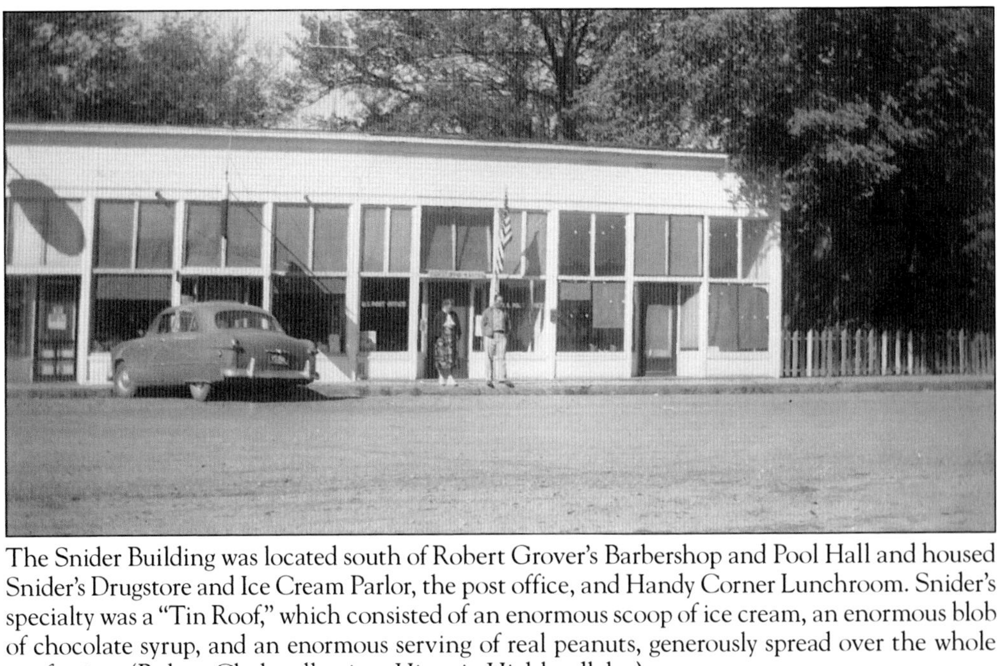

The Snider Building was located south of Robert Grover's Barbershop and Pool Hall and housed Snider's Drugstore and Ice Cream Parlor, the post office, and Handy Corner Lunchroom. Snider's specialty was a "Tin Roof," which consisted of an enormous scoop of ice cream, an enormous blob of chocolate syrup, and an enormous serving of real peanuts, generously spread over the whole confection. (Robert Clark collection, Historic Highlandlake.)

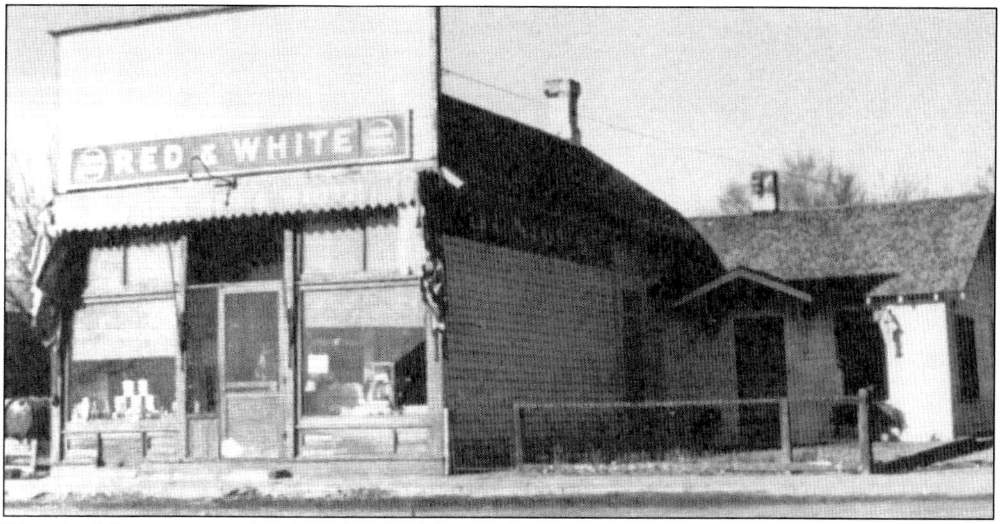

This building is one of only a few buildings physically moved from Highlandlake to Mead that is still standing. Sometime in the late 1920s or early 1930s, a grocery chain called the Red & White moved in. After the grocery store closed in the 1950s, the building became the Mead Post Office until the current post office was built in 2001. (Mead Bulldog Annual collection, Historic Highlandlake.)

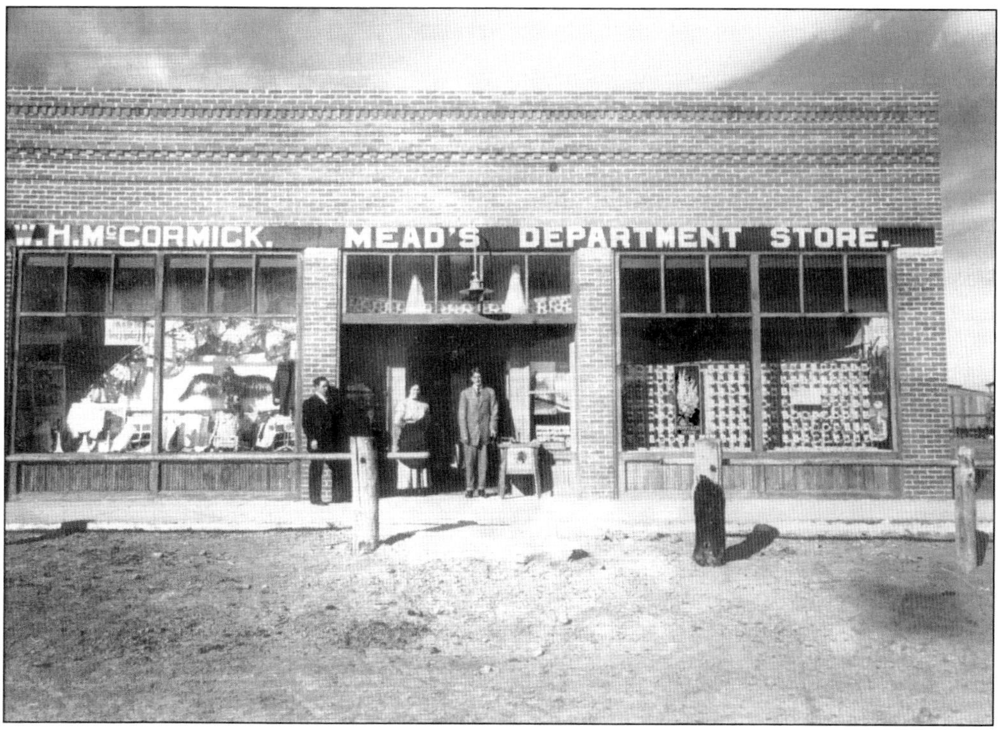

W.H. McCormick of Berthoud, a man of many talents, opened several stores in northern Colorado, including one in Mead. General merchandise, including dry goods such as fabrics and clothing, paper products, seeds and garden implements, and everything else in between, were sold. In a sense, stores like this were the forerunners of the one-stop stores of today. (Kelly Jorgensen collection, Historic Highlandlake.)

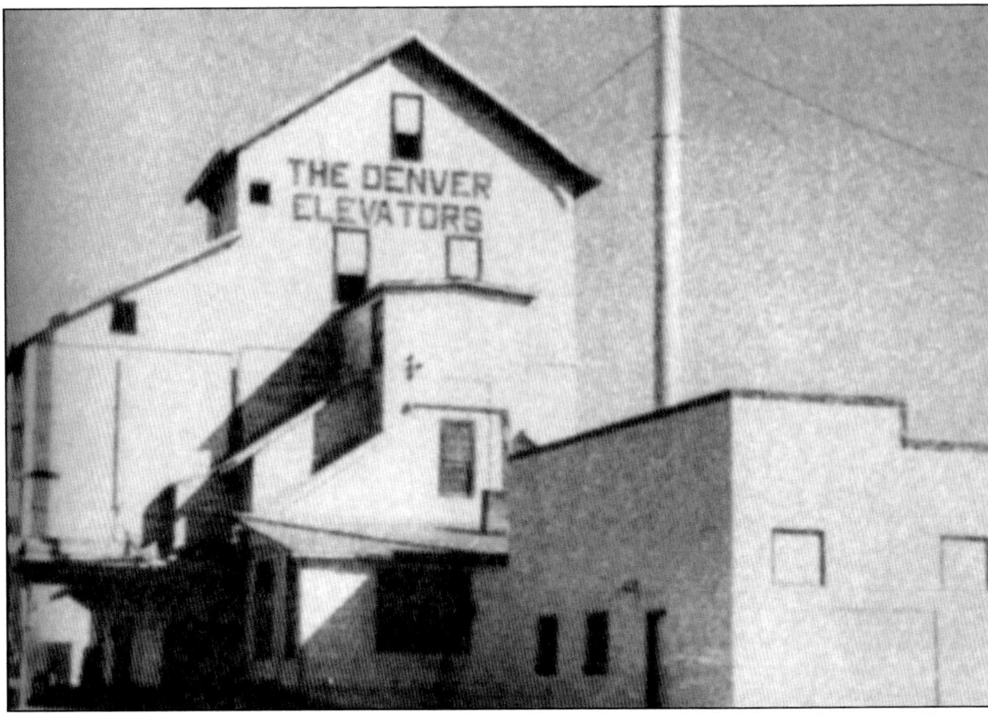

Within a year or two of the town of Mead's founding, there were already three elevators. The two owned by a company called Denver Elevator are still standing. The third, Mead Farmer's Elevator, was located at the south end of Main Street. It was dismantled in the late 1950s and reassembled on a lot near Longmont. (Mead Bulldog Annual collection, Historic Highlandlake.)

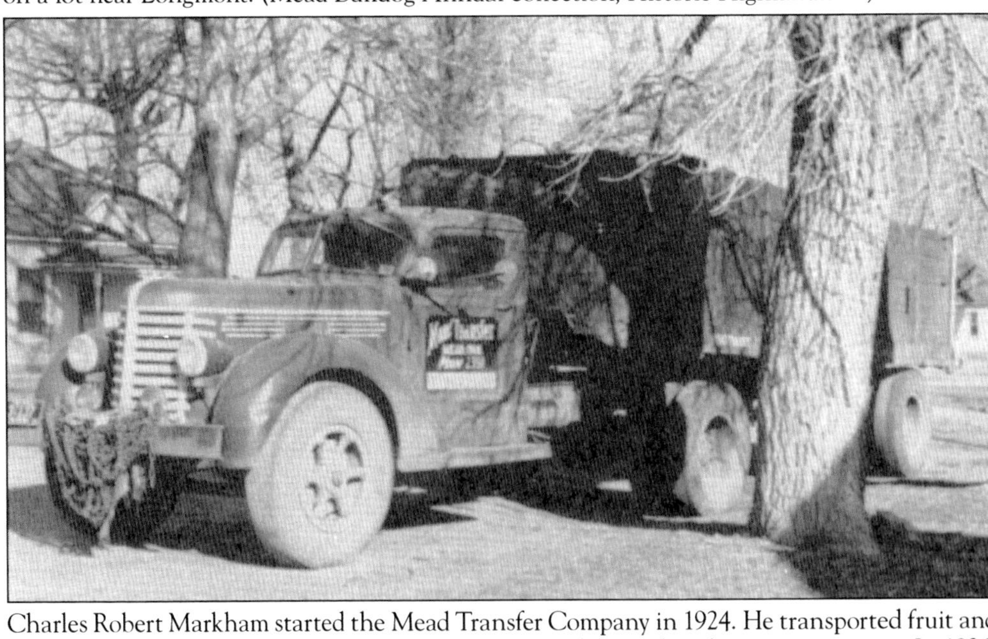

Charles Robert Markham started the Mead Transfer Company in 1924. He transported fruit and produce from the Western Slope and in the off-season, delivered coal to area customers. In 1936, Charles had a stroke, after which his son Rodney Markham took over the business until 1954. (Mead Bulldog Annual collection, Historic Highlandlake.)

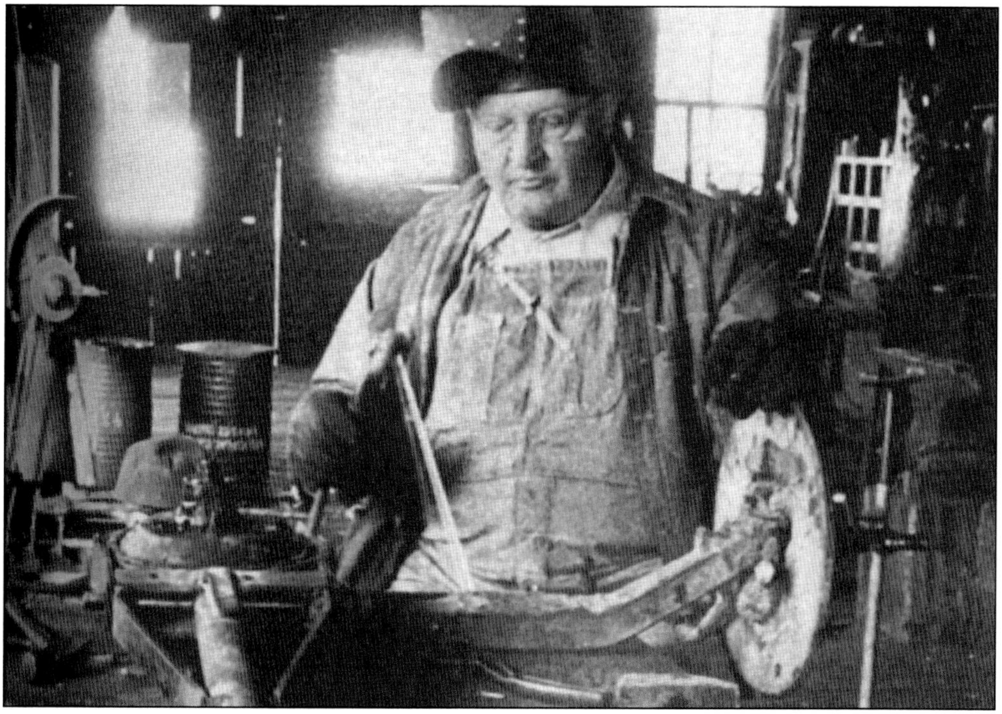

Francis Michael "Frank" Boro was one of a succession of blacksmiths working out of a building located across the road from the Ford dealership. Later in life, he contracted throat cancer and used a device to help him speak. That did not stop him from working. This photograph, taken in the early 1950s, shows Boro hard at work welding. He died in 1955. (Mead Bulldog Annual collection, Historic Highlandlake.)

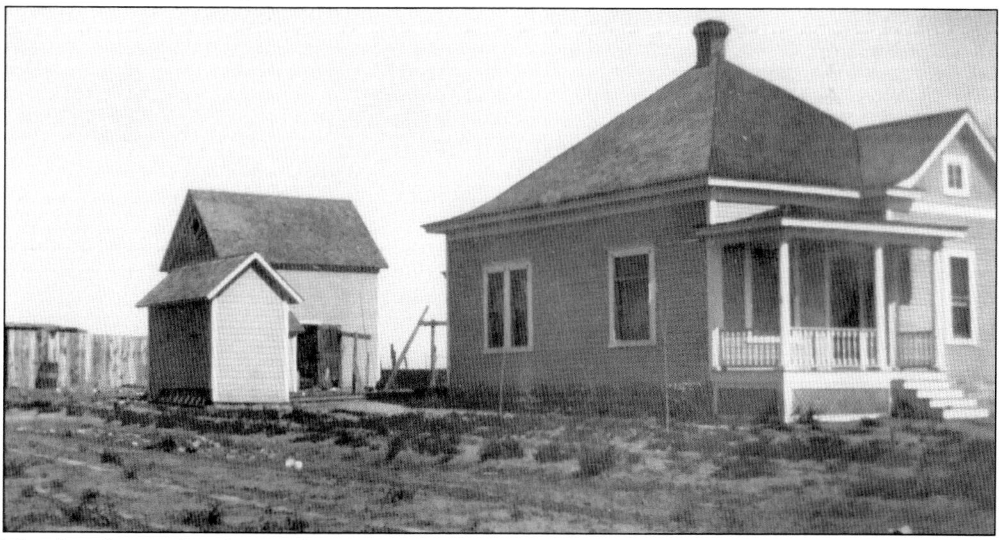

The first few weeks after Paul Mead drew up his town plat were very busy. By February 19, 1906, when he finally got around to traveling to Greeley and officially filing, he had already sold over 75 lots. Most of early buildings contained businesses, but before long, private homes started springing up. In 1911, Charles Goodwin built this home on Fifth Street. (Linda and Paul Gehring collection, Historic Highlandlake.)

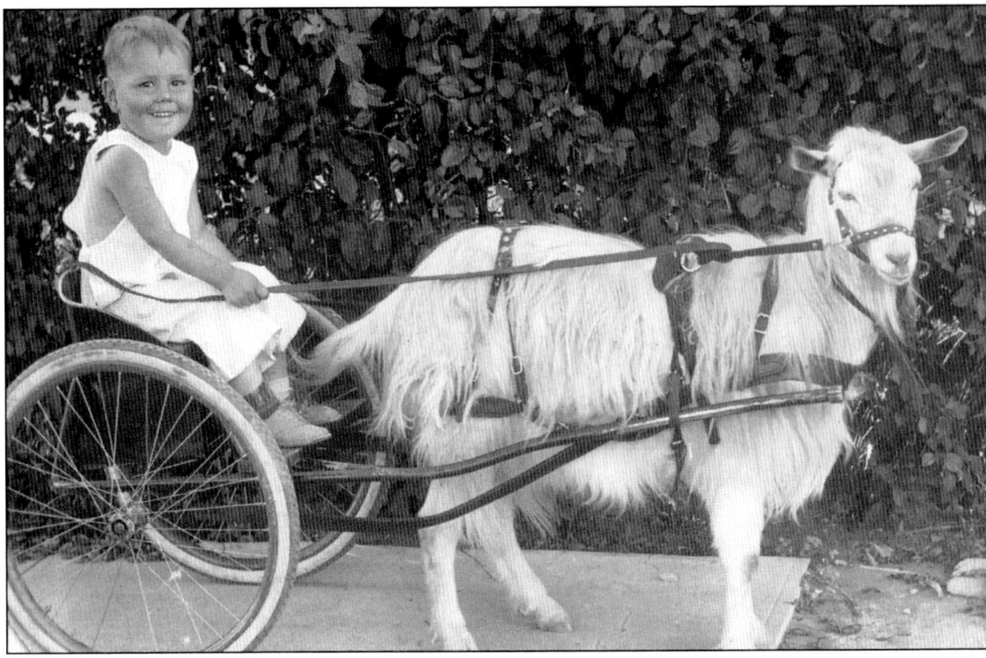

During the first half of the 20th century, itinerant photographers, often with a pony or goat in tow, traveled through small towns and villages offering to take photographs of children. Pictured is Harley Bruce Markham, son of Harley Bruce and Ruth (McCormick) Markham. This photograph was taken around 1935. (Hester Markham collection, Historic Highlandlake.)

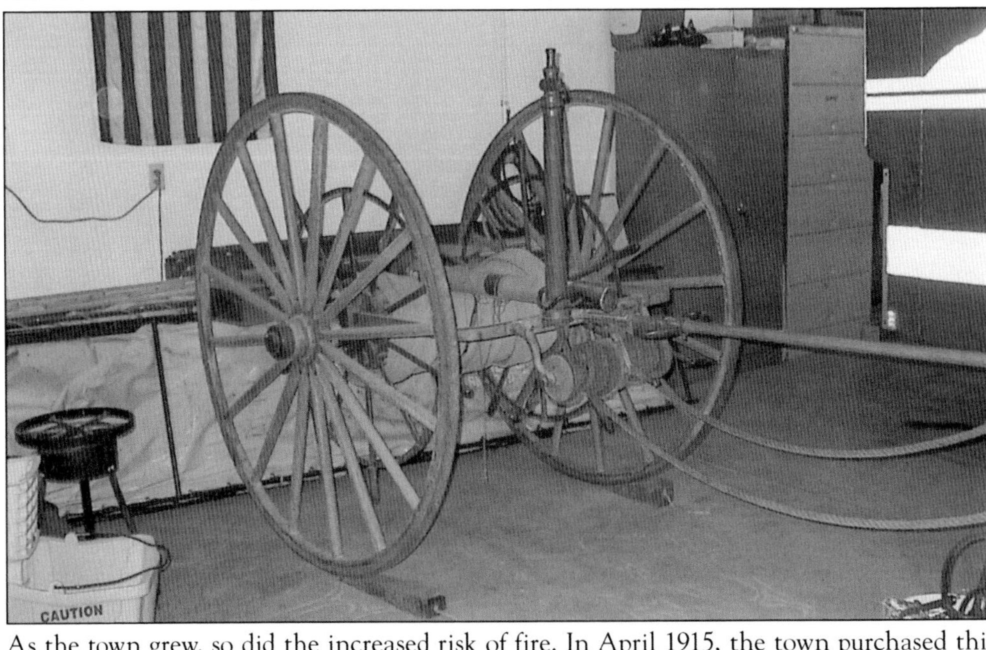

As the town grew, so did the increased risk of fire. In April 1915, the town purchased this human-pulled hose wagon from the Eureka Fire Hose Company of New York. Firefighters harnessed themselves to the cart and raced to the fire. To keep in practice, teams from all over the Front Range would complete in hose cart races each summer. (Pauli Driver Smith collection, Historic Highlandlake.)

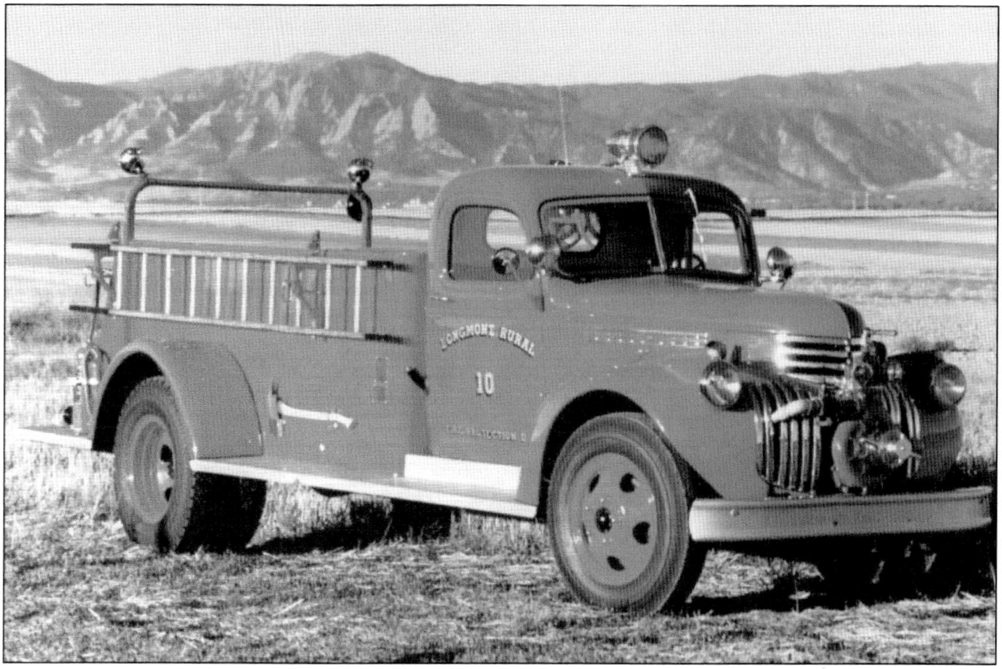

In 1960, a year before the town joined the Mountain View Fire Protection District, Mead purchased a used 1947 Chevrolet truck from Longmont Rural Fire Department. This was the first motorized fire engine owned by Mead, replacing the hand-drawn hose cart used since 1915. Engine No. 10 was sold in 1968 to the unincorporated town of Niwot. (Mountain View Fire Protection District collection, Historic Highlandlake.)

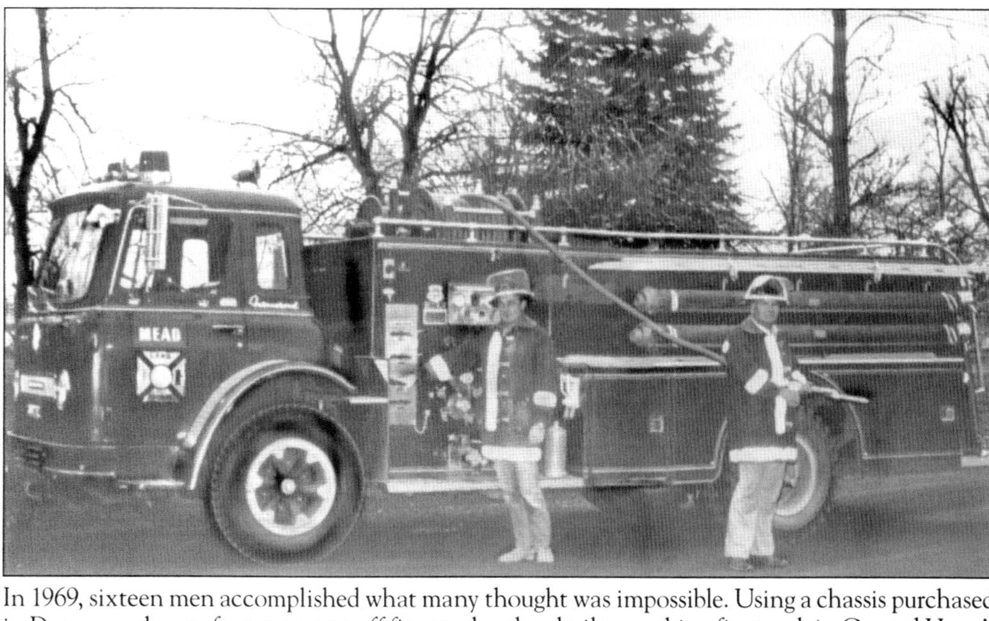

In 1969, sixteen men accomplished what many thought was impossible. Using a chassis purchased in Denver and parts from two cast-off fire trucks, they built a working fire truck in Conrad Hopp's farm shop. Conrad's son Craig, a student at Longmont High School, machined some parts in the school's shop. The total cost was about $2,000. Those pictured are Richard "Dick" Cornell, left, and Melvin Lebsack, right. (Dick Cornell collection, Historic Highlandlake.)

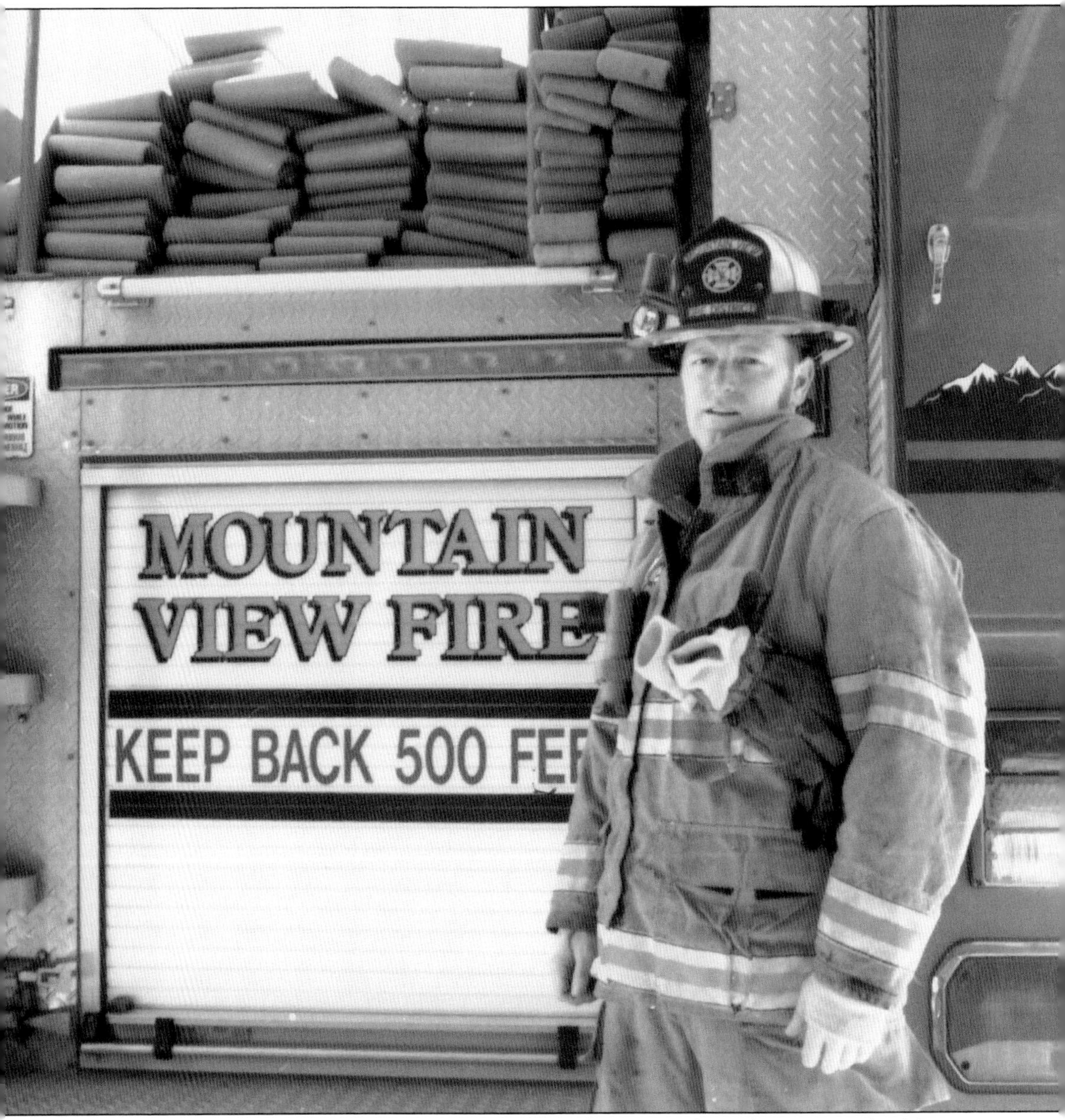

As part of a photograph documentary exhibit for the Town of Mead Centennial celebration in 2006, Sheila Koenig photographed firefighter Christopher Clark standing beside one of Mountain View Fire Protection District's trucks. Mead was part of a group of area towns and rural areas that formed this district in 1961. Until recently, Mead was one of the last remaining all-volunteer departments in Colorado. (Sheila Koenig.)

Three

Farming and Irrigation

With the exception of a few small spring-fed prairie potholes or buffalo wallows, the area around what eventually became Highlandlake was semiarid. The soil was rich but, without a reliable water supply, farmers considered most of the land uncultivable. Lorin C. Mead, the first settler in this area, originally thought that the prairie pothole he named Highland Lake was sufficient for his needs; however, the drought of 1873 quickly changed his mind. To save his struggling wheat, it was imperative to finish the 25-mile-long Highland Ditch from Lyons to Highland Lake. With almost 19 miles of ditch still to go, Lorin C. Mead resolved to do the impossible—build the remainder of the ditch in one day.

People are still debating whether Mead's firsthand account, published on March 15, 1889, in the *Longmont Ledger*, detailing how he accomplished this amazing feat, is 100-percent fact or even physically possible. However, there is no denying that thanks to his determination, powers of persuasion, and hard work, thousands of acres of once dry land are now under irrigation.

With a reliable water supply established, new settlers quickly flocked to the area. Besides wheat, barley, hay, peas, potatoes, and later corn and sugar beets, most farms around the lake had fruit orchards and a variety of berry crops. Nearly everyone raised animals as well, including cattle, sheep, goats, rabbits, and poultry.

Today, after decades of gentle slumbering, development is coming back to the Highlandlake and Mead area. The town of Mead, along with the entire southwest corner of Weld County, is booming. As development continues to grow, it is placing increasing pressure on the surrounding rural areas for shared resources such as land and water. Just as it was in Lorin C. Mead's time, water holds the key to the area's future. In the coming decades, how the competing interests work to compromise and learn to share this increasingly finite resource will most certainly determine the future of this beautiful area.

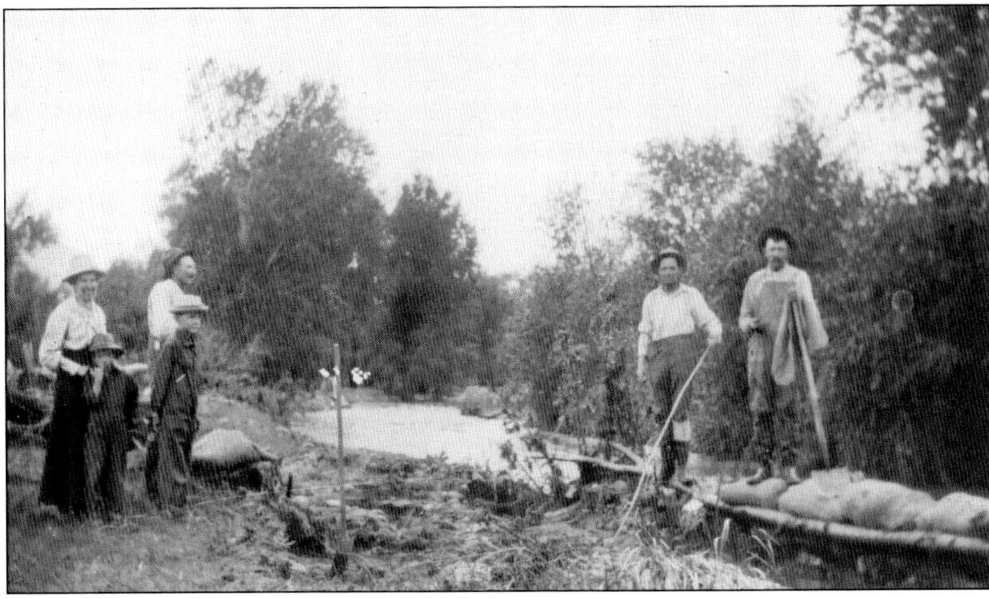

The Highland Ditch was vital to the success of the Highlandlake community and the surrounding farmland. Ditches take work to maintain, they need to be cleared of debris and, in this case, sandbagged when running too full. Mrs. Lorin C. (Elizabeth) Mead and two young sons, Rufus, left, and Malcolm, right, are standing on the left bank. The men are unidentified. (Mary Mead Jensen collection, Historic Highlandlake.)

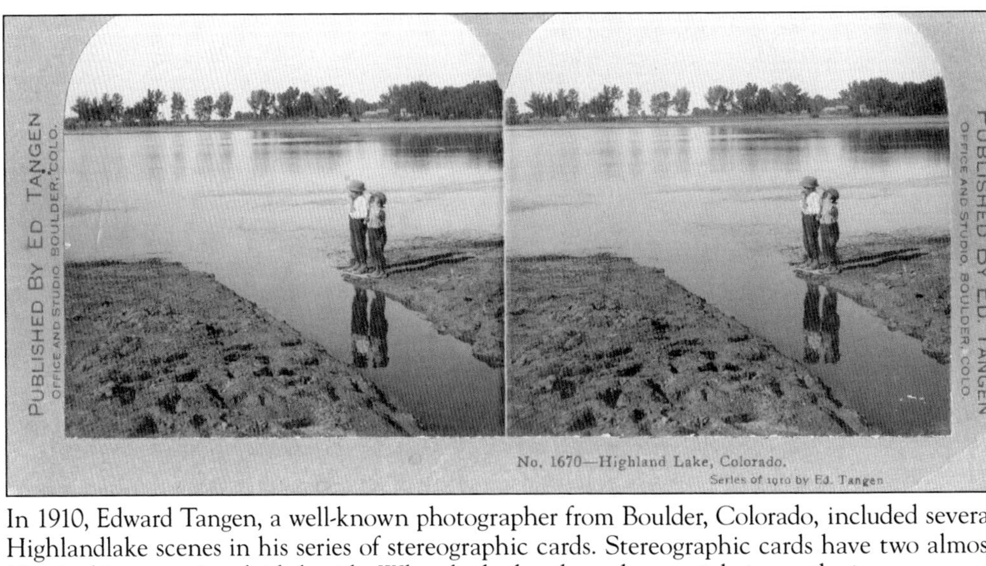

In 1910, Edward Tangen, a well-known photographer from Boulder, Colorado, included several Highlandlake scenes in his series of stereographic cards. Stereographic cards have two almost identical images printed side by side. When looked at through a special viewer, the image appears to be three-dimensional. In this photograph, two young boys stand on the west side of the inlet ditch to Highland Lake. (Mark French collection, Historic Highlandlake.)

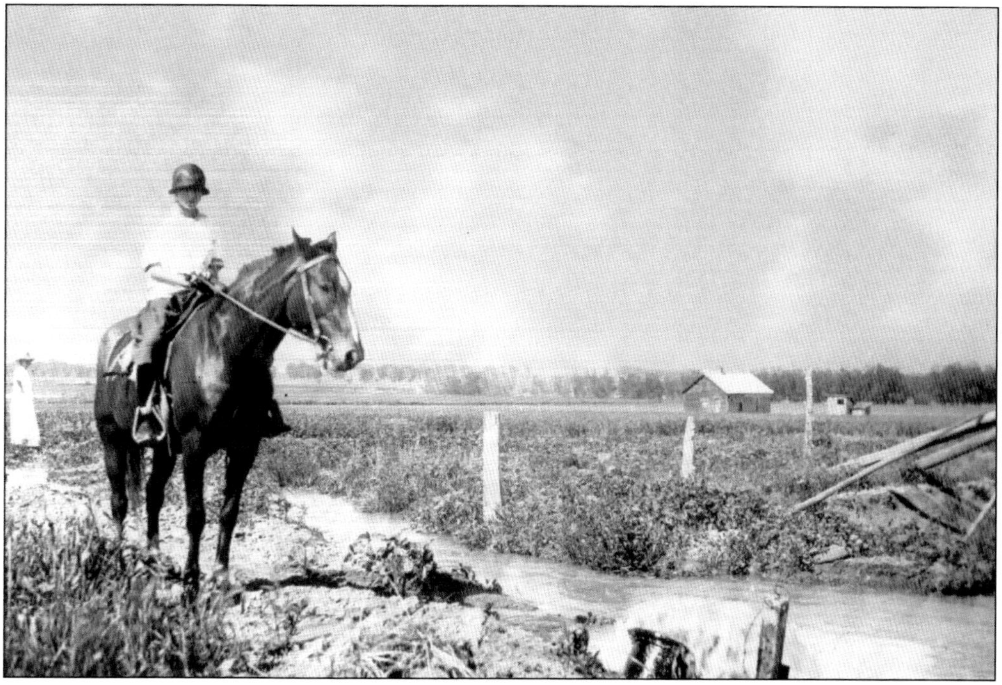

Western women were not exempt from working in the fields. As a young widow, Harriett Markham managed to raise a large family and run her Locust Grove Farm, located on the northern edge of the old Highlandlake District. In 1917, her daughter Elizabeth "Bessie" Markham checks the irrigation water from horseback while another woman, possibly her mother, watches in the background. (Glen and Patricia Markham collection, Historic Highlandlake.)

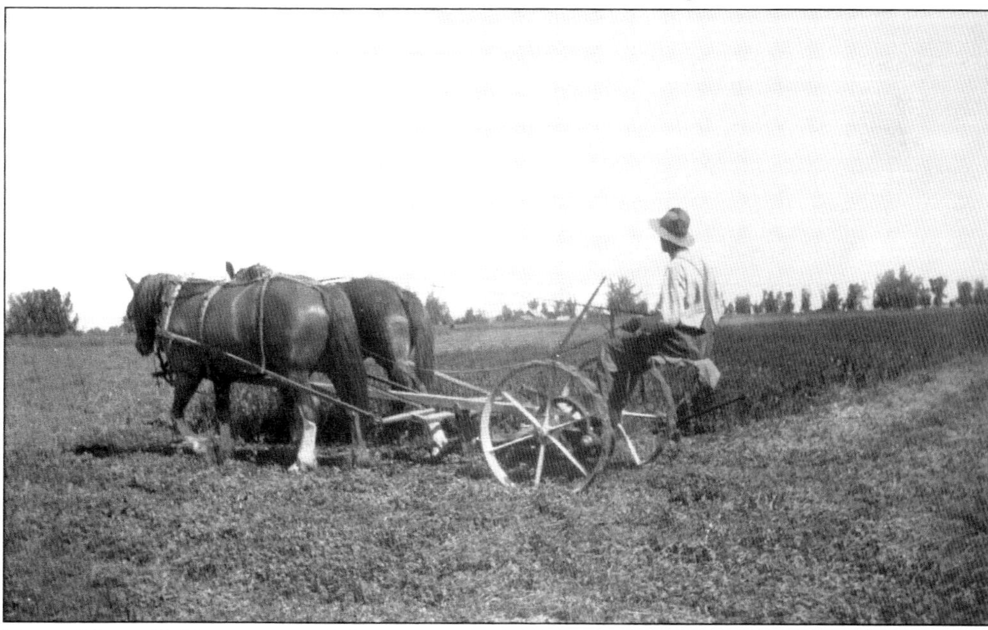

Meent Harm Habbinga, son of German immigrants, cuts hay with a single-tree horse-drawn mower. It took a horse going at a steady, moderate pace about 2.5 minutes to cut a .25-mile-long-by-3.5-foot-wide swath. (Longmont Museum and Cultural Center.)

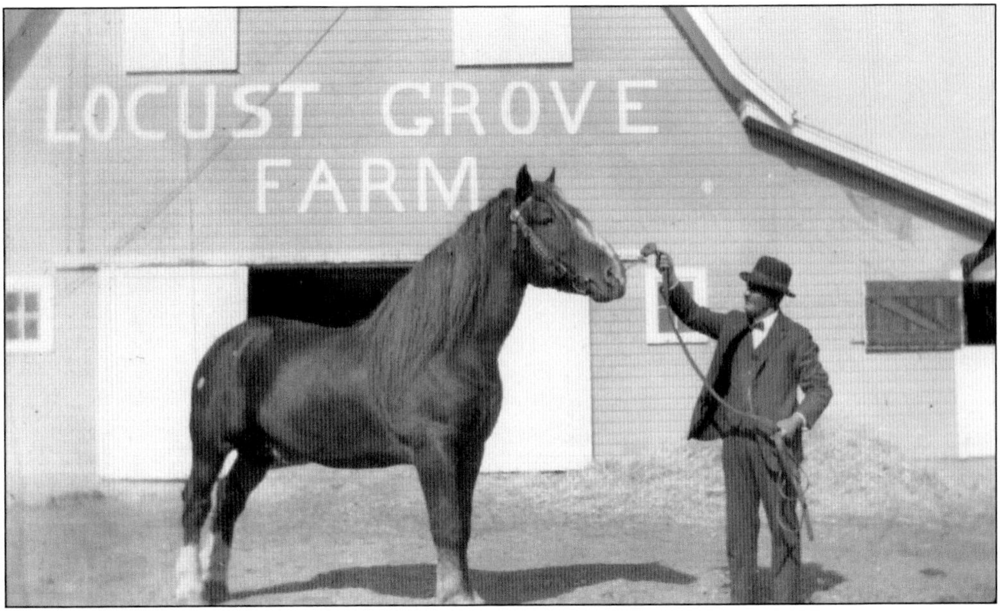

In 1917, Edmond Markham shows off one of his prize horses. Until tractors came into wide use, horses and mules were very important on a farm. They provided the power that pulled most of the heavy farm machinery, including plows, hay rakes, and wagons. Even the early steam-powered portable engines were usually built on skids or wheels and pulled to the work area with horses. Shown below, a year later, three teams of horses made stacking hay on the Markham's Locust Grove Farm easy—especially compared to having to stack the hay by hand. (Both, Glen and Patricia Markham collection, Historic Highlandlake.)

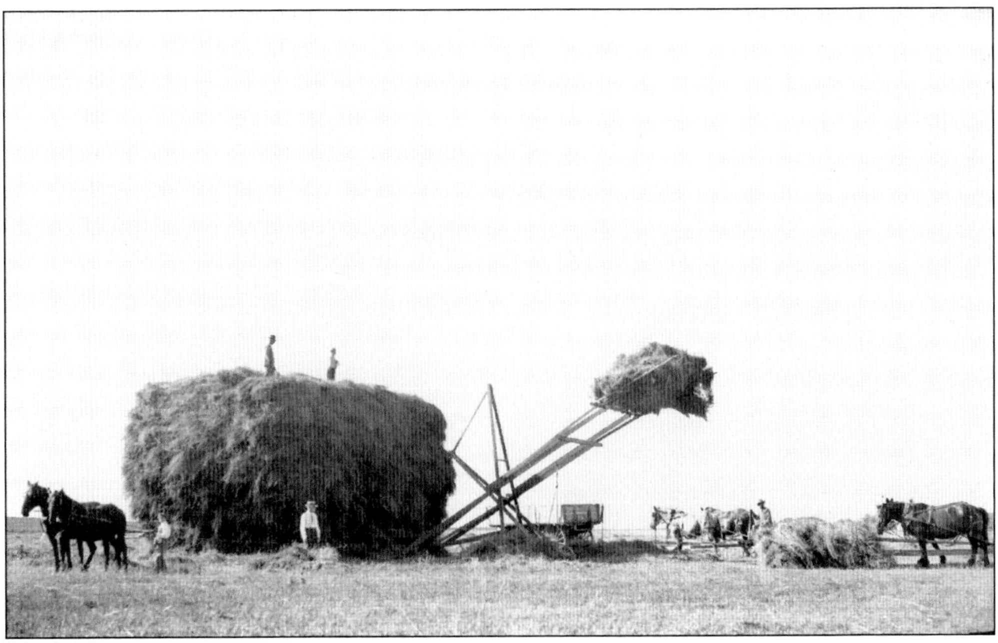

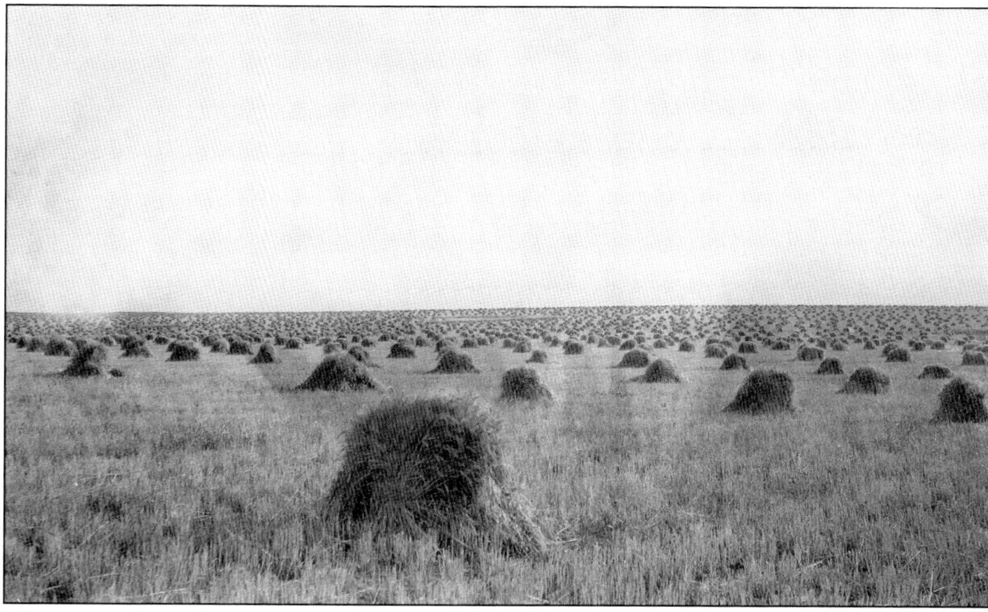

Wheat was the earliest crop raised in the Highlandlake area. It was an ideal and profitable crop since once planted, except for occasional irrigating, it could generally be left alone until harvest, thus allowing a farmer to focus on other things. One mile west of Mead, sheaves of winter wheat are stacked in a field to dry until the threshers come. (Longmont Museum and Cultural Center.)

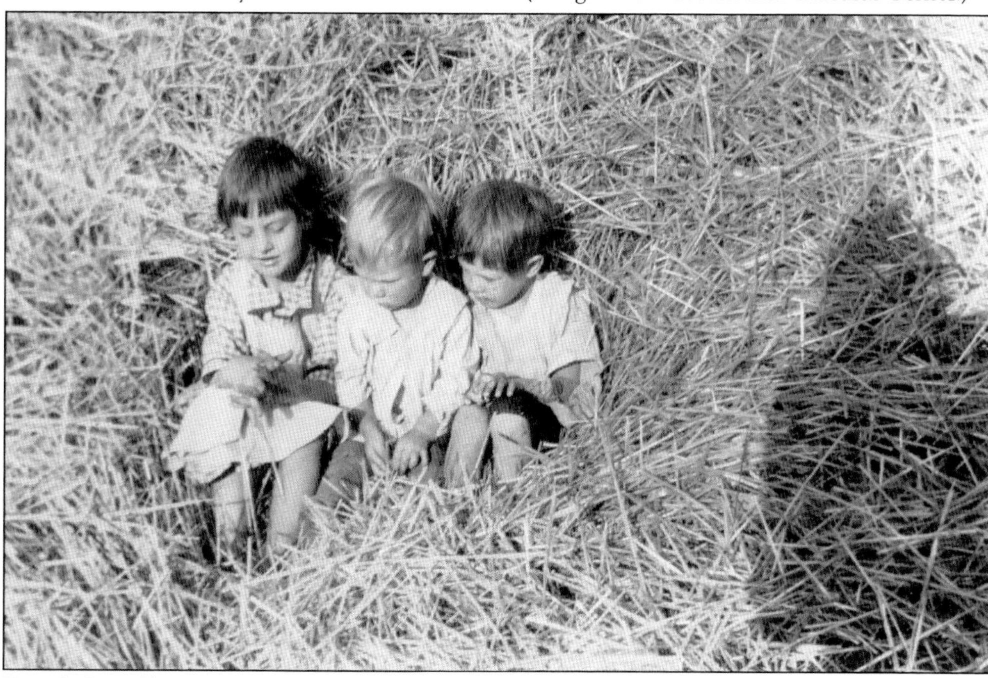

Farm kids and haystacks seem to go together. In this c. 1925 photograph, from left to right, Margaret Marie, Earnest, and Florence, children of Mark and Gertrude "Trudy" Kiteley of Liberty, talk over their plans for the day. Liberty was located at the intersection of Colorado Highway 66 and Weld County Road 5. All that is left are a few homes and the old Liberty Hall Grange. (Nancy Kiteley collection, Historic Highlandlake.)

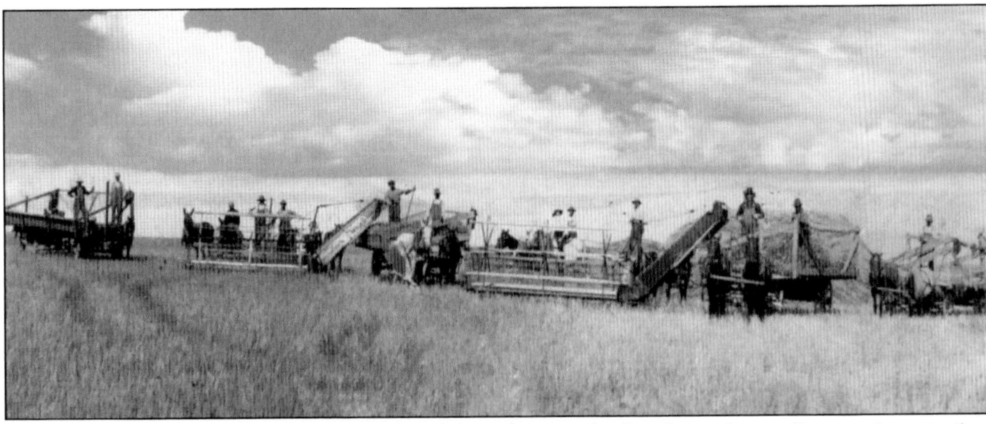

Threshing, which separates the kernels of the wheat or barley from the stalk, was done in late summer or early fall. The equipment was expensive and took many people to operate, so usually groups of farmers went together to form threshing companies and then traveled from farm to farm until all the threshing was finished. In 1887, Highlandlaker George Davis and eight other men formed a threshing syndicate, with F.P. Woodley as the manager. The community quickly dubbed the new enterprise "The Bald-Headed Company." Over the years, other companies formed. This panoramic photograph of an unidentified threshing crew was probably taken on the Markham's Locust Grove Farm around 1910. Shown below, Bill Kiteley's farm was located next to the Liberty highline beet dump. (Above, Hester Markham collection, Historic Highlandlake; below, Nancy Kiteley collection, Historic Highlandlake.)

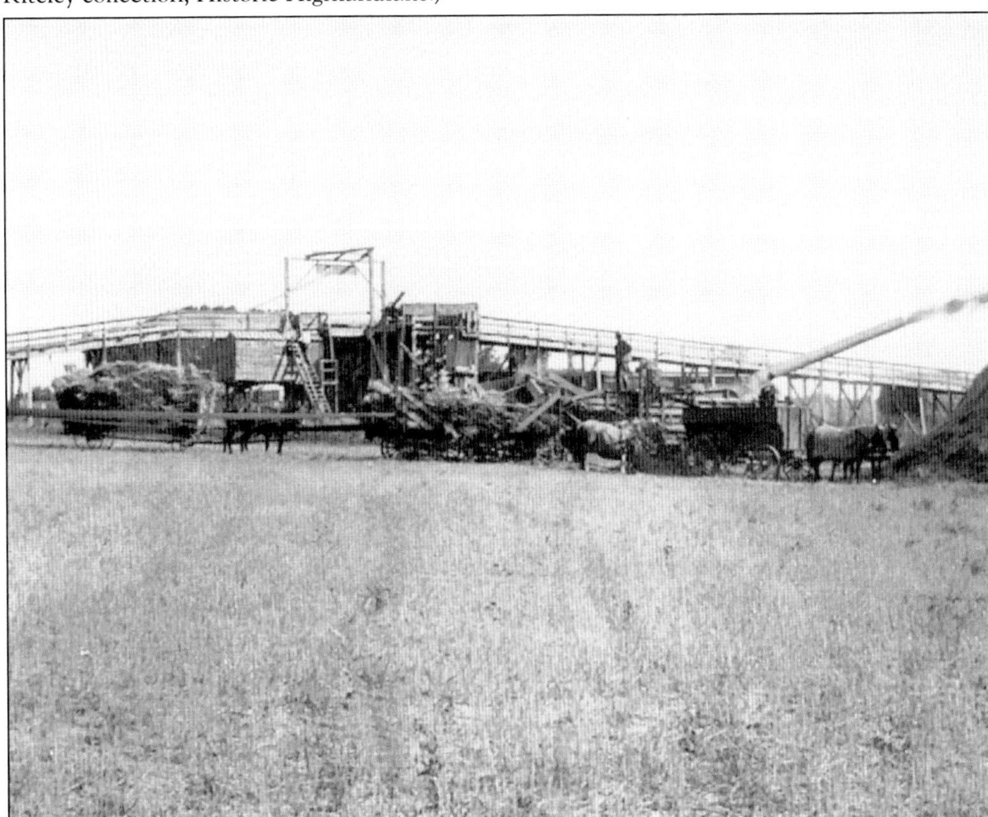

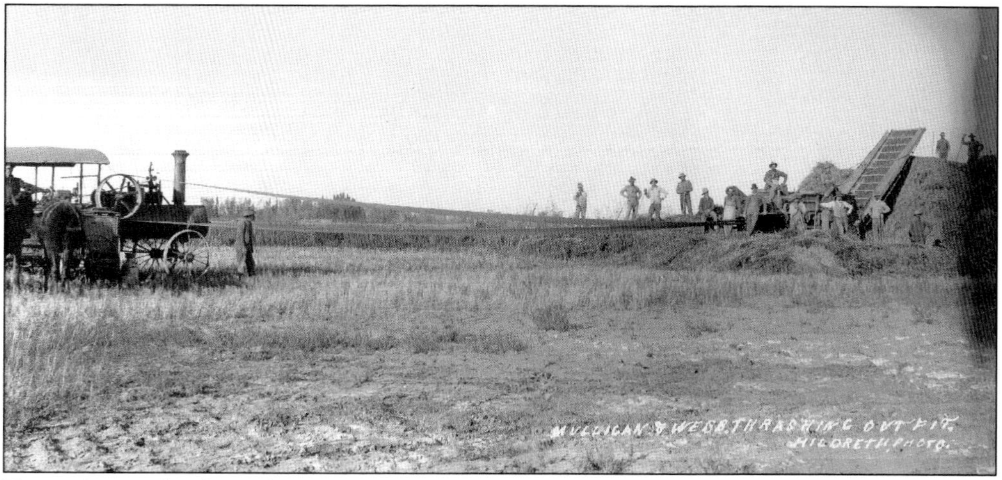

Two miles south of Mead in the community of Mount Zion, farmers Billy Mulligan and Ben Webb also formed a threshing company. By the early 1900s, steam-powered engines came into wider use. They were more efficient than the old horse-drawn threshers, but took more people to run since someone needed to constantly shovel coal into the firebox. (Longmont Museum and Cultural Center.)

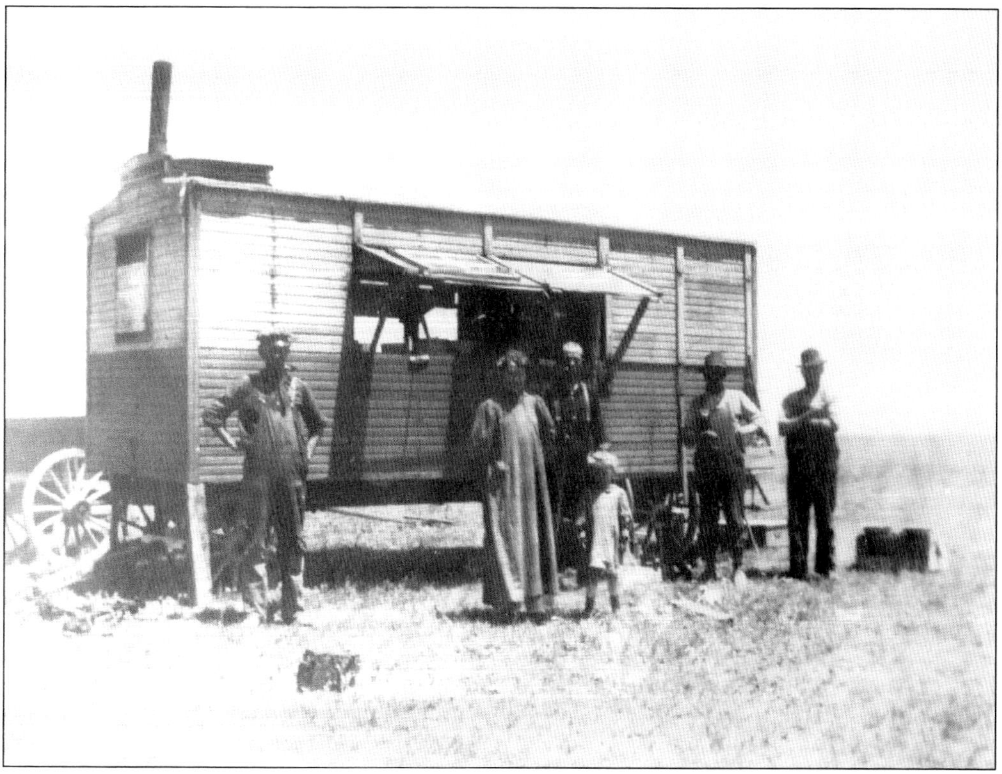

Threshing crews traveled to wherever the harvest took them. In the summer and fall of 1911, Malcolm Mead traveled all the way to Wyoming, taking his family with him. His wife, Florence (at center), cooked for the crew. Malcolm is standing next to her. The little girl in front is their daughter Dorothy. The chuck or cook wagon also doubled as the family's sleeping quarters. (Mary Mead Jensen family collection, Historic Highlandlake.)

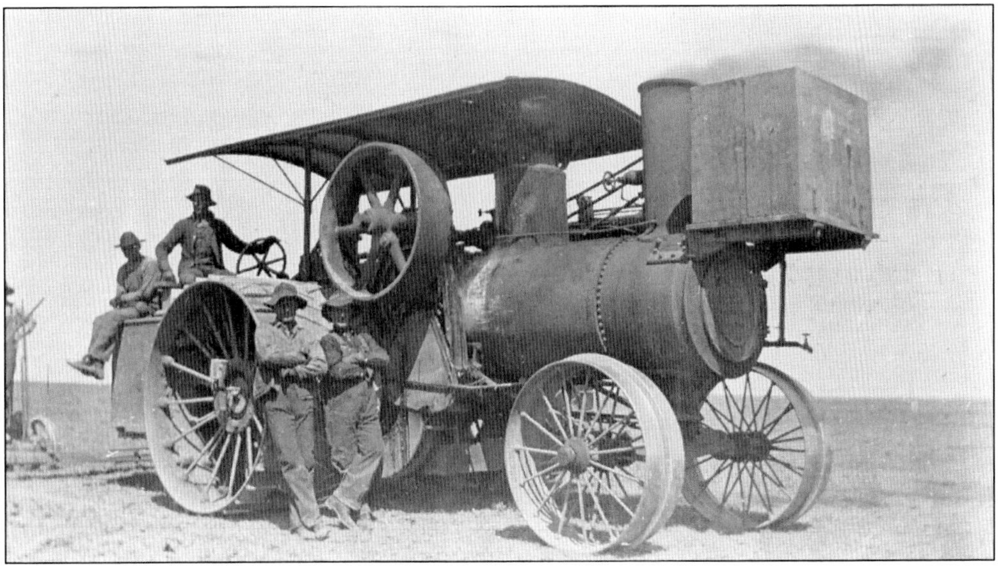

While threshing kept Malcolm Mead and his crew busy in late summer and fall, they were not above taking whatever work came their way. If they were not threshing, they might be found plowing fields or helping to harvest sugar beets. However, the work was hard and often dangerous. In the September 19, 1911, *Longmont Ledger*, the Highlandlake column reports that "Otto Leonard had the misfortune to have one of his arms hurt so badly in the threshing machine while at work on Will Brown's farm, that it was necessary to have it amputated at the Longmont hospital last week." (Both, Mary Mead Jensen family collection, Historic Highlandlake.)

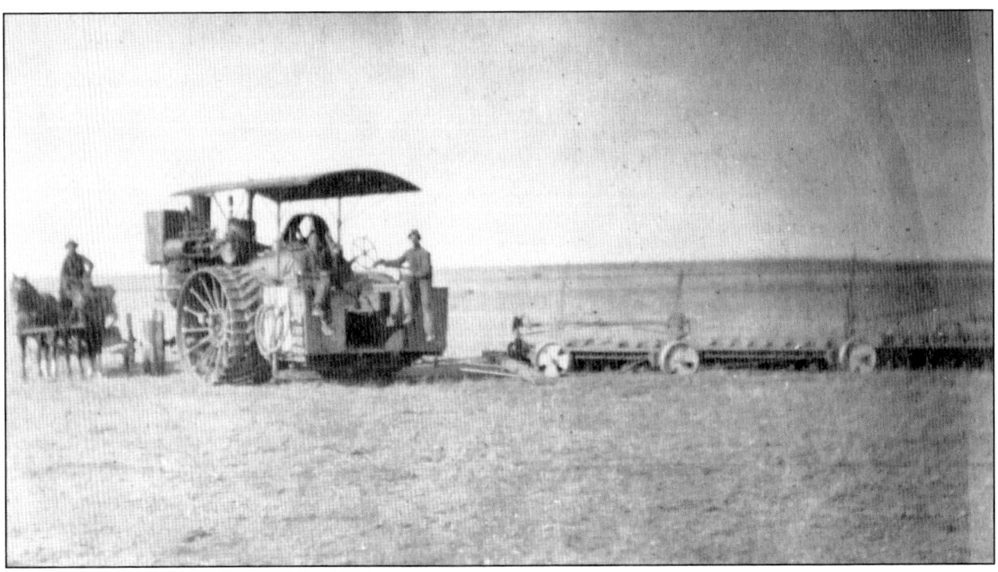

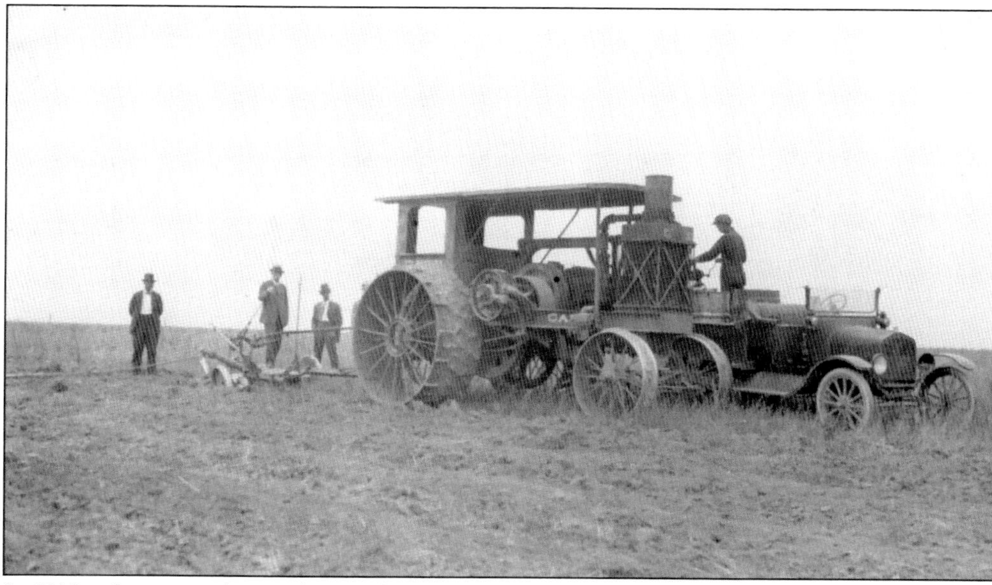

In 1918, when considering the purchase of a "traction engine," farmers could choose between steam or gasoline. Edmond Markham chose a Case gasoline tractor for his Locust Grove Farm. The young man standing in the car next to the tractor is filling it with gas. (Glen and Patricia Markham collection, Historic Highlandlake.)

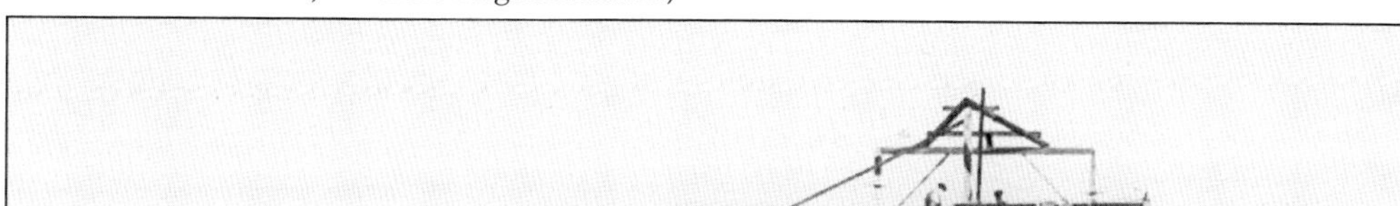

The Great Western Railway created several beet dumps along its Johnstown-to-Longmont route. The Liberty dump was located on Colorado Highway 66 halfway between present-day Weld County Roads 5 and 7. Horses pulled wagons up the ramp and dumped the beets into waiting railroad cars. Leonard Litzenberger often reminisced about how as a young boy, he thought that taking a horse-drawn wagon over the beet dump was "the next best thing to a roller coaster ride." (Nancy Kiteley collection, Historic Highlandlake.)

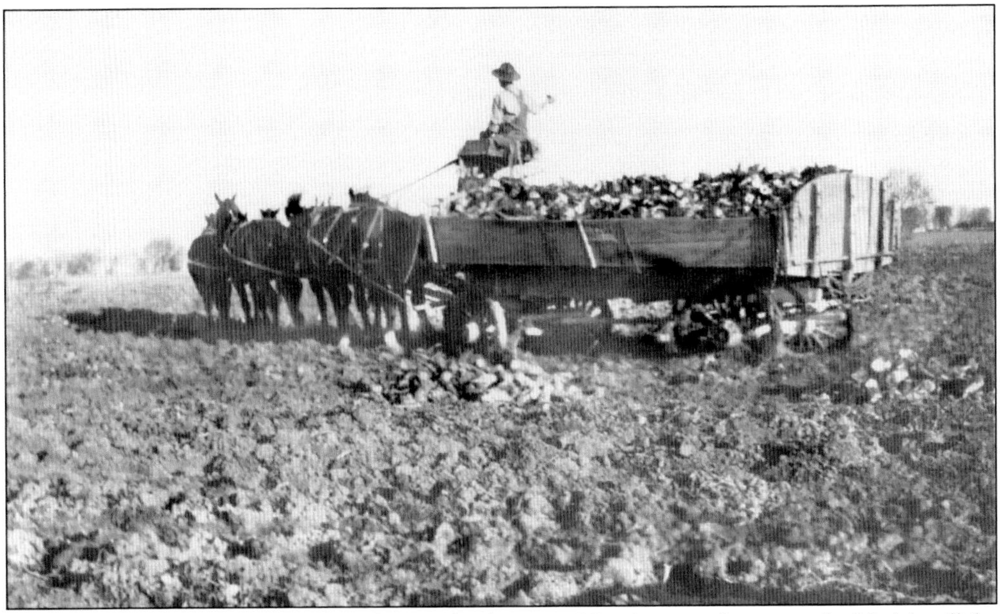

By 1918, sugar beets were a major cash crop for this area. Nearly every farmer in the Highlandlake and Mead area had most, if not, all of his acres planted in beets. Harvest usually occurred in early to mid-October. Here, Bill Kiteley hauls a wagonload of beets to the Liberty beet dump. (Nancy Kiteley collection, Historic Highlandlake.)

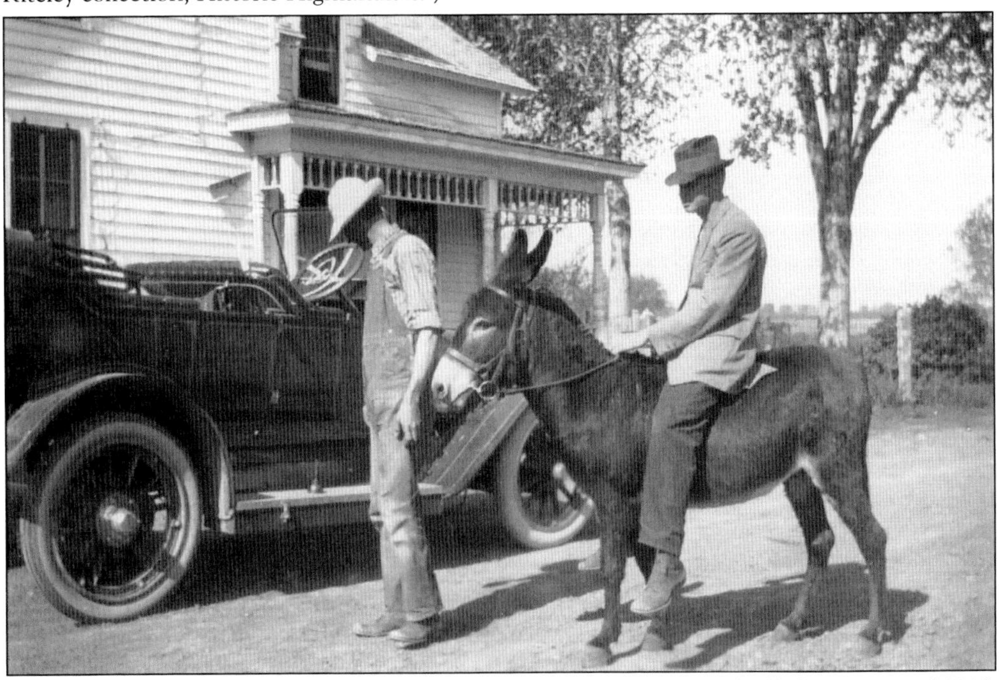

Brothers Bill (left) and Mark Kiteley strike a humorous pose in front of Bill's home around 1915, highlighting the contrast between old and new modes of transportation. By this time, the whole country was at a transportation crossroads. Cars and tractors frequently traversed Highlandlake and Mead area roads and farms, and both could be purchased at dealerships in Mead. (Nancy Kiteley collection, Historic Highlandlake.)

Tom and Ernest "Ernie," sons of Mark and Gertrude (Spencer) Kiteley, inspect their new seed corn elevator around 1940. The Kiteley family of the Liberty-Mead area was in the foreground of the development of a hybrid corn seed that could successfully grow in the short northern Colorado growing season. (Nancy Kiteley collection, Historic Highlandlake.)

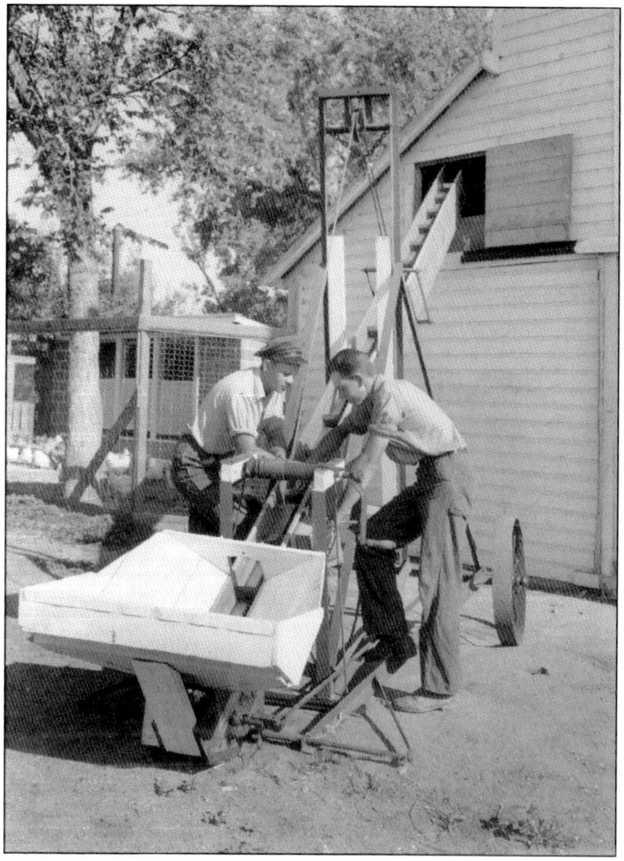

When viewed across the span of the 20th century, mechanization has had a profound effect on farm productivity. For example, at the end of the 19th century, it took 35 to 40 hours of planting and harvesting labor to produce 100 bushels of corn. A hundred years later, the same amount of corn takes Edward "Bud" Minch just a little over two hours. (Pauli Driver Smith.)

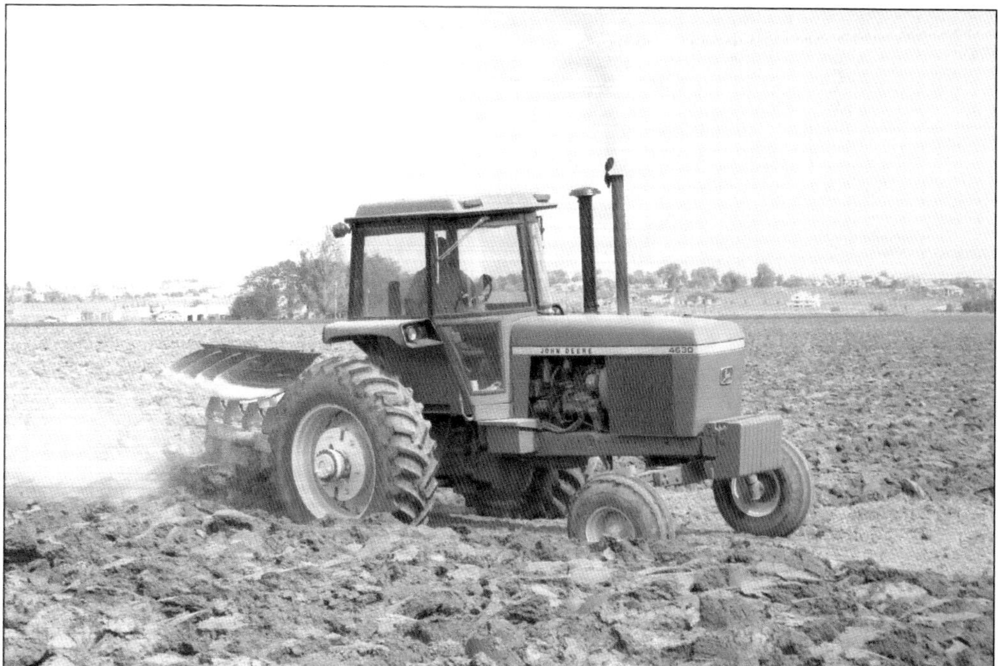

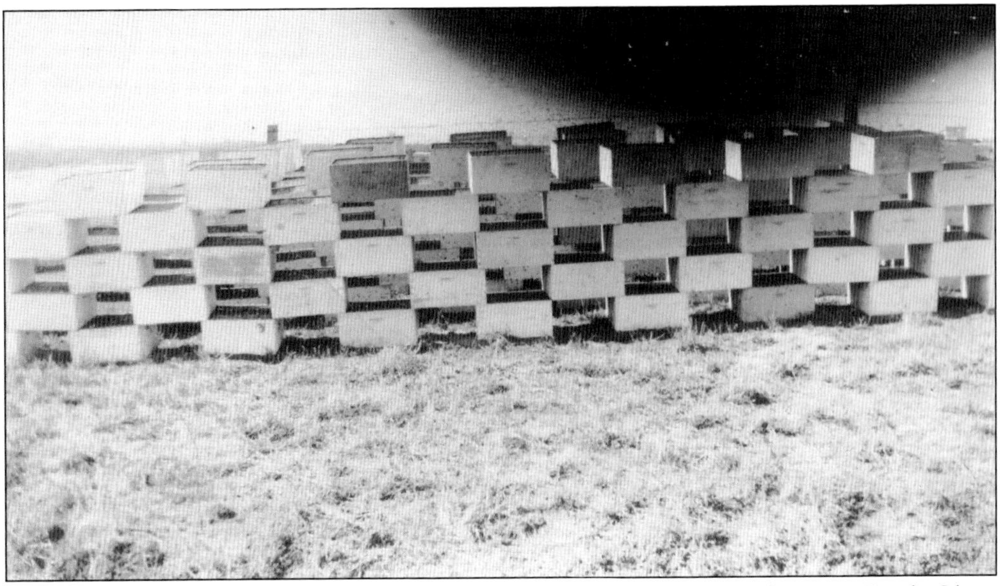

During the 1930s and 1940s, Swedish immigrant Ivar Olson kept bees on his sister Emely Olson Eckman's Highlandlake farm. What appears to be a large stain in the upper-right area is actually a huge swarm of bees. Emely's home was the former Congregational Church of Highlandlake's parsonage. (Jean Eckman Nelson collection, Historic Highlandlake.)

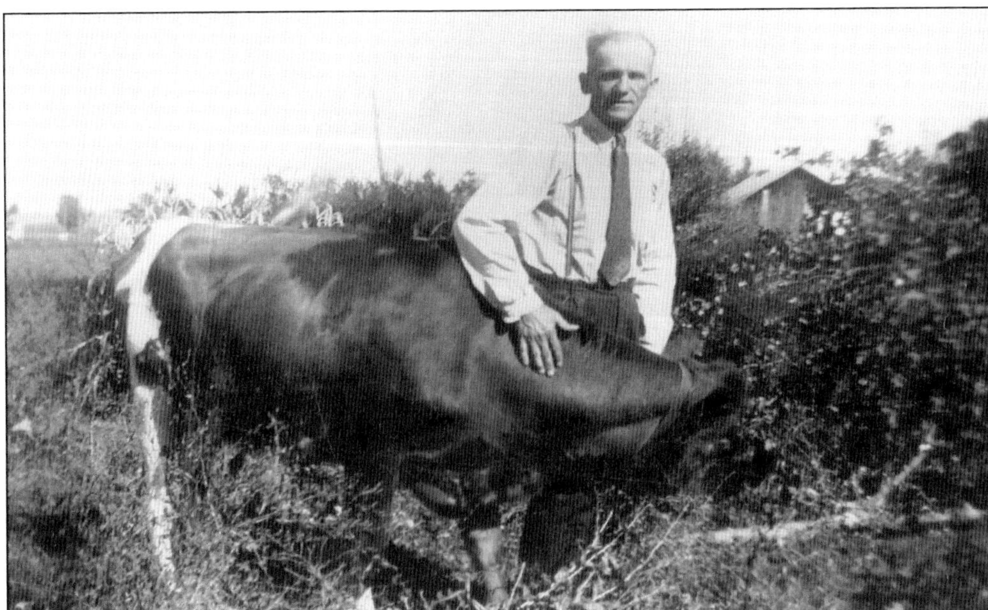

Farming was not just confined to rural areas; even townies often kept poultry, a milk cow or two, and of course, until cars became common, at least one horse to ride or pull a buggy. In 1942, Mead resident James Elmer Newton shows off Bessie, one of his beloved cows. (Patricia Newton French collection, Historic Highlandlake.)

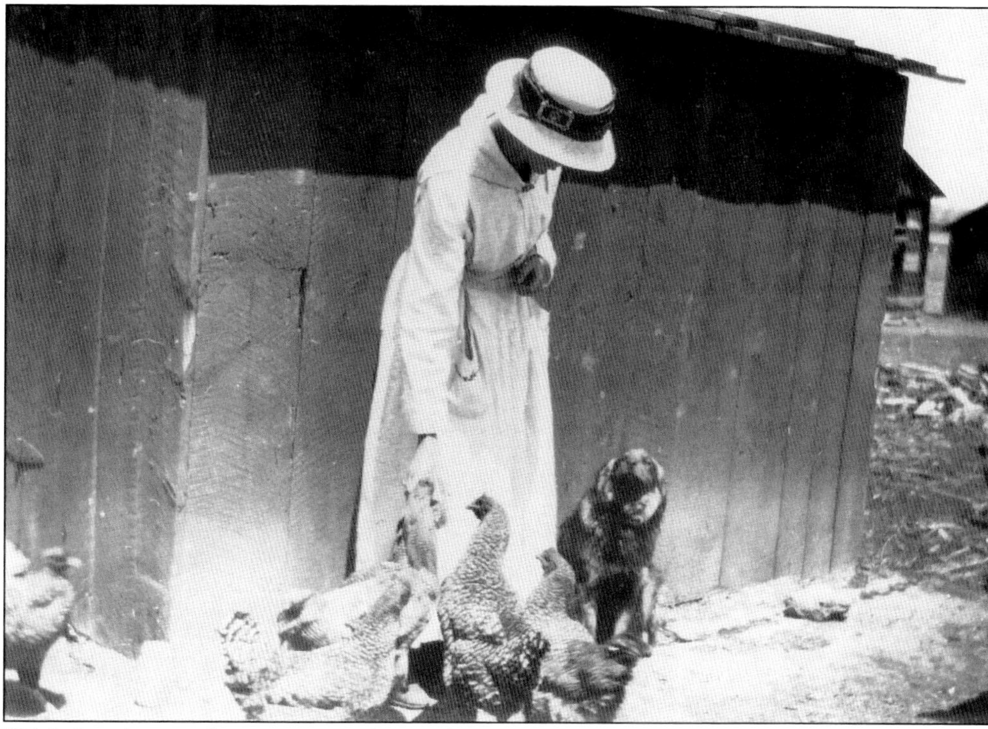

While her dog watches on, Gertrude "Trudy" Spencer Kiteley, wife of Mark K. Kiteley, tends to her first flock of hens on their Liberty farm in 1917. Chickens were a staple on most farms and in town. They produced eggs for breakfast and baking and meat for the dinner table. (Nancy Kiteley collection, Historic Highlandlake.)

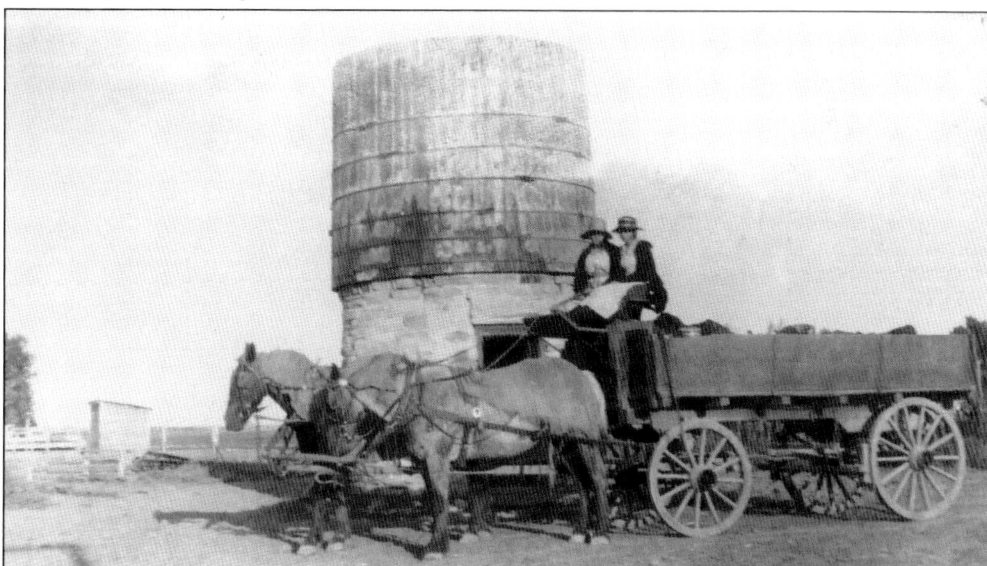

Two of Harriett Markham's daughters drive a loaded wagon past their Locust Grove Farm water tower. Water towers were common as they provided water storage for people and animals alike. Old burlap sacks are draped across the backs and necks of the horses to protect them from flies and other flying pests. (Glen and Patricia Markham collection, Historic Highlandlake.)

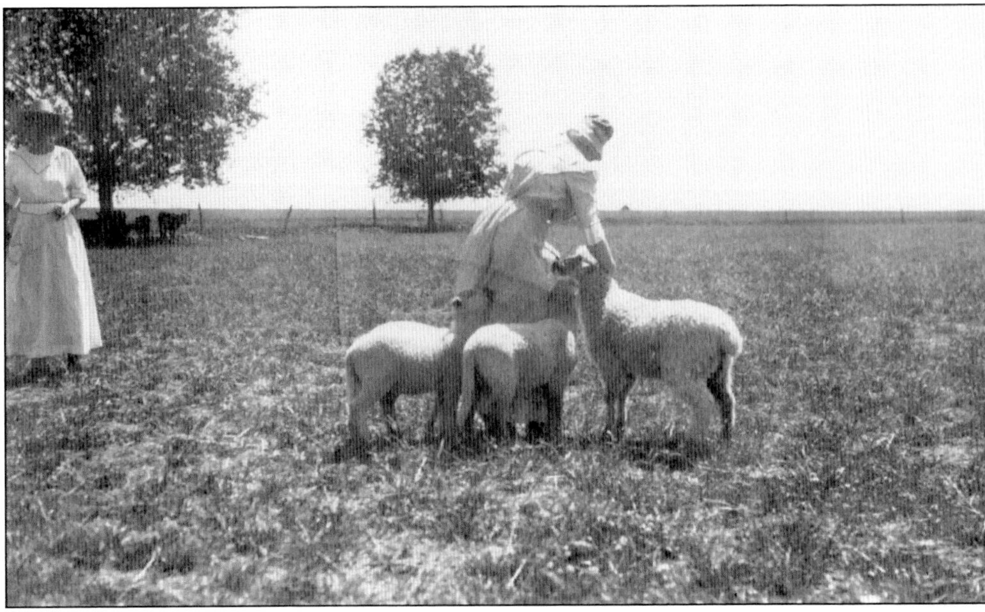

Besides field crops, the Markham family also raised Hampshire sheep and cows on their Locust Grove Farm. In the first of these rather idyllic photographs, taken in 1917, a woman appears to be feeding a lamb while two younger lambs impatiently wait their turn. In the photograph below, three women, a man, and two young children stand in the background with a few lambs while the photographer tries to take in the whole scene. It appears that the sheep in the foreground are drinking water from a flood-irrigated field while the calves hover close to the fence line far away from the action. (Both, Glen and Patricia Markham collection, Historic Highlandlake.)

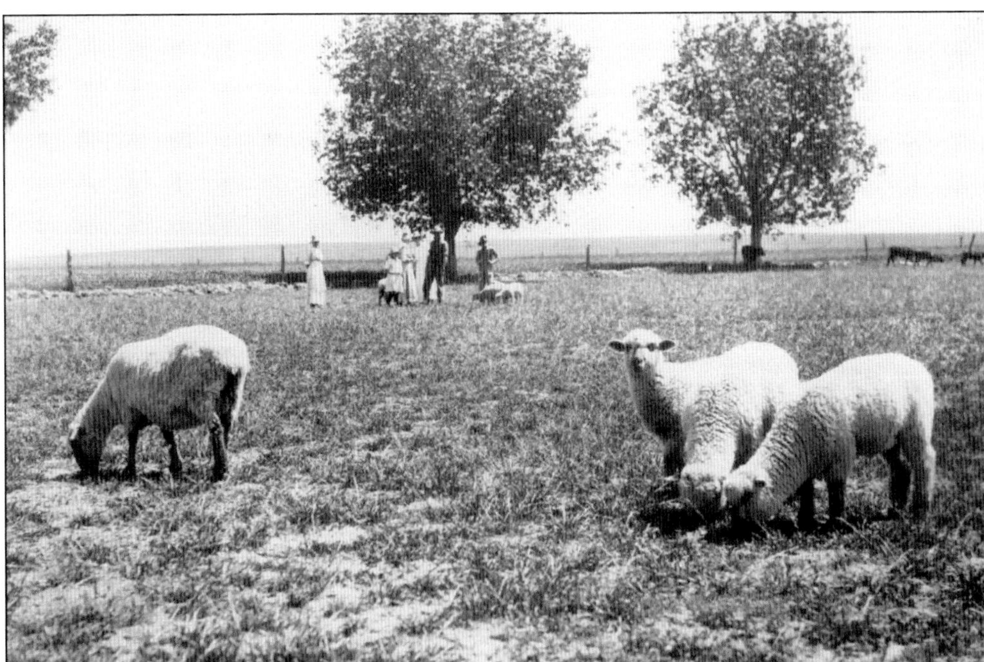

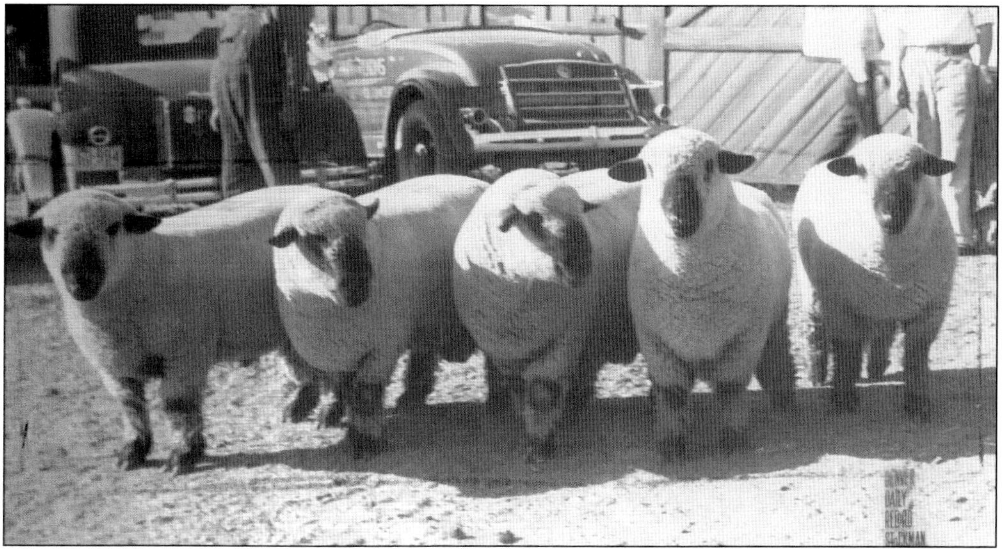

The National Western Stock Show, founded in 1906 and held each January in Denver, is the largest stock show in the world. By 1928, when Edmond Markham exhibited these prizewinning Hampshire lambs, prices, which were at 11¢ per pound, were the highest on record. The market crashed the following year, and prices did not recover for more than a decade. (Glen and Patricia Markham collection, Historic Highlandlake.)

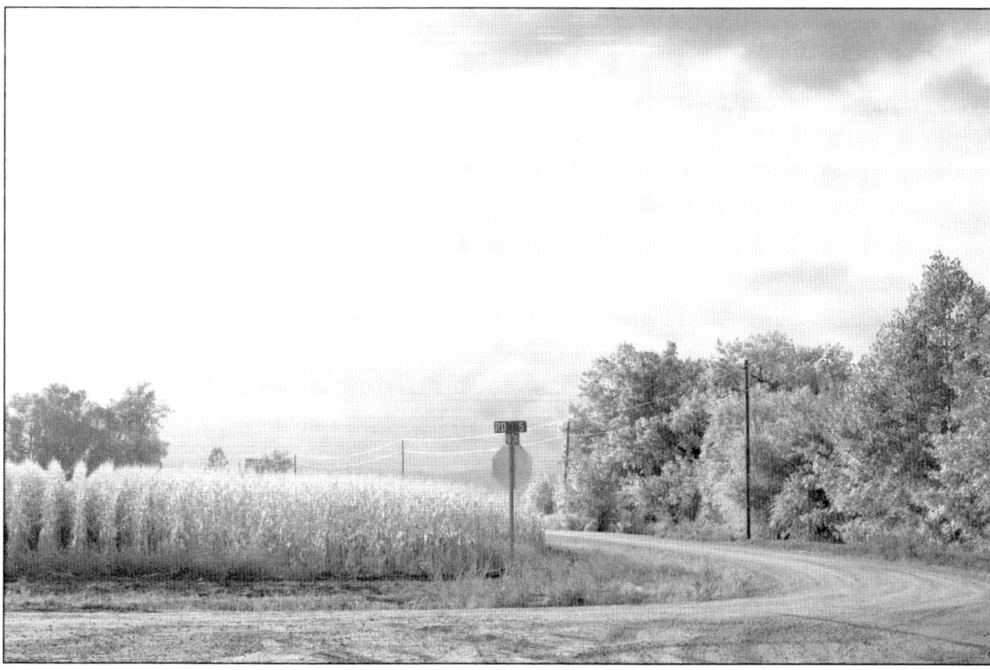

Fall is in the air as corn dries in a field near the southwest corner of Highland Lake. A storm is brewing in the distance hiding most of the mountains from view. A large two-story home once sat on this corner and housed the first post office in the Highlandlake district. (Pauli Driver Smith.)

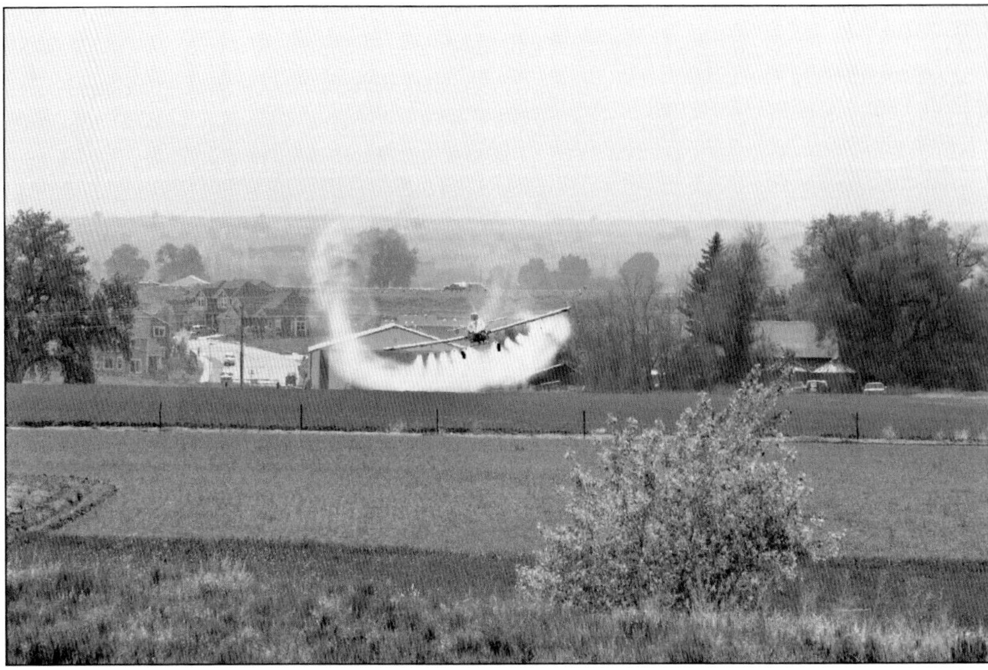

A crop duster flies low over a field just west of the North Creek subdivision in Mead and then swoops under power lines and over a cornfield southwest of Highlandlake. Since the early days of aviation, farmers have utilized aircraft to dust fields, help monitor irrigation systems, and check livestock and the development of crops. The International Flying Farmers got its start in 1944, when several farmers who owned airplanes met in Stillwater, Oklahoma, with the idea of organizing a club. Within a few years, the group spread across the United States and eventually to other countries. (Both, Pauli Driver Smith.)

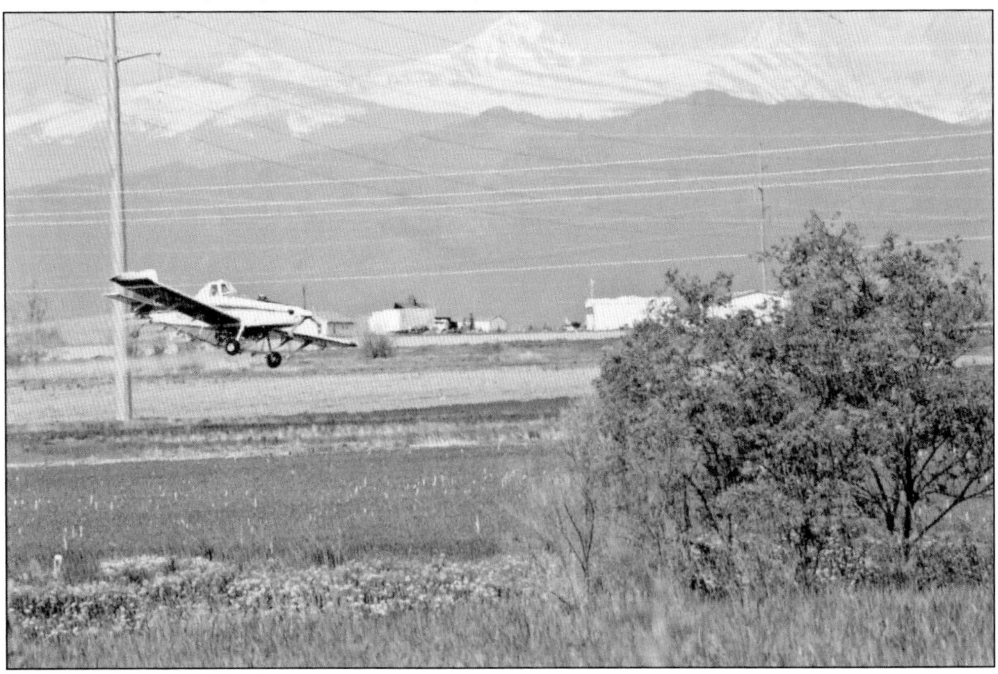

Four

Reading, Writing, and Arithmetic

The Highlandlake School had its beginnings in 1877, when Lorin C. Mead called a meeting to discuss applying for a school district. The first year, students attended classes in the downstairs of the Hubbell home, with daughter Emma Hubbell as their teacher. In the fall of 1878, a one-room schoolhouse, located near the southeast corner of the lake, was finished. Over the years, additions to the building increased classroom space. In 1910, the school consolidated with the Mead School District, although students still attended school in Highlandlake for several more years. The Highlandlake School closed for good in December 1920, marking the end of an eventful era. When the children returned to school in January, they attended the Mead Consolidated Schools. Eventually, the school district plowed most of the schoolhouse into a pit and burned it. Fortunately, the Brossman family saved the original one-room section and moved it up the street to their home. In 1995, Steve and Cheryl Senesac converted the former schoolhouse into a preschool, thus bringing the building full circle.

When Paul Martin Mead laid out his new town, he set aside land for a two-room school, found just north from what is now the Mountain View Fire and Town Hall Building. Consolidation, a growing movement to help make local school districts economically viable, was taking hold in Weld County, and Mead was no exception. As small schools in the surrounding area, such as Highlandlake, Pearl Howlett, and Pleasant View, merged into the Mead District, the student population swelled. In 1917, a large brick building, made to house grades one through twelve, was built on Welker Avenue between Main (Fourth Street) and Fifth Streets. The school opened in the fall of 1918 and closed in the early 1960s, when the district merged into the St. Vrain Valley School District. The school district eventually razed all but the gym, built in the 1930s by the Works Progress Administration (WPA). A new elementary school, built around the old gym, is now standing on the site of the old schoolhouse.

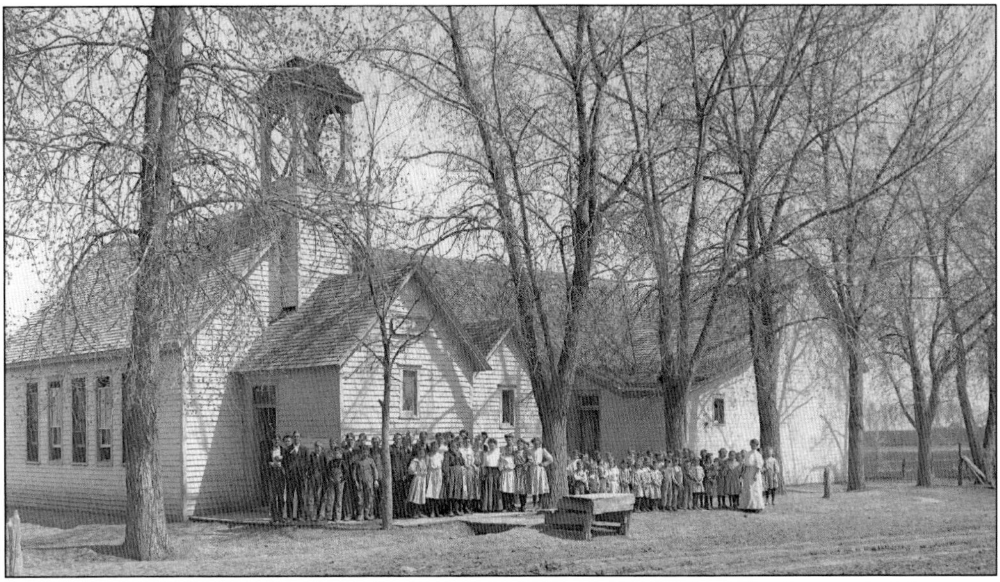

In 1906, when Helen Strong started the first grade in the Highlandlake School, several additions to the original one-room building accommodated a growing population of students. In 1887, there were only 43 students. By 1900, the student population grew to almost 100, and the school experimented with adding high school classes. After only two years and to the dismay of the community, the high school teacher, Professor Billings, left for a better position. Once again, students wanting to extend their education past the eighth grade had to travel eight miles to Longmont or board with family or friends in Greeley. In the photograph below, there is a wooden pole laid across the crotches of three trees. The swings are wrapped tightly around the pole. (Above, Mary Ellen Litzenberger collection; below, Hester Markham collection, Historic Highlandlake.)

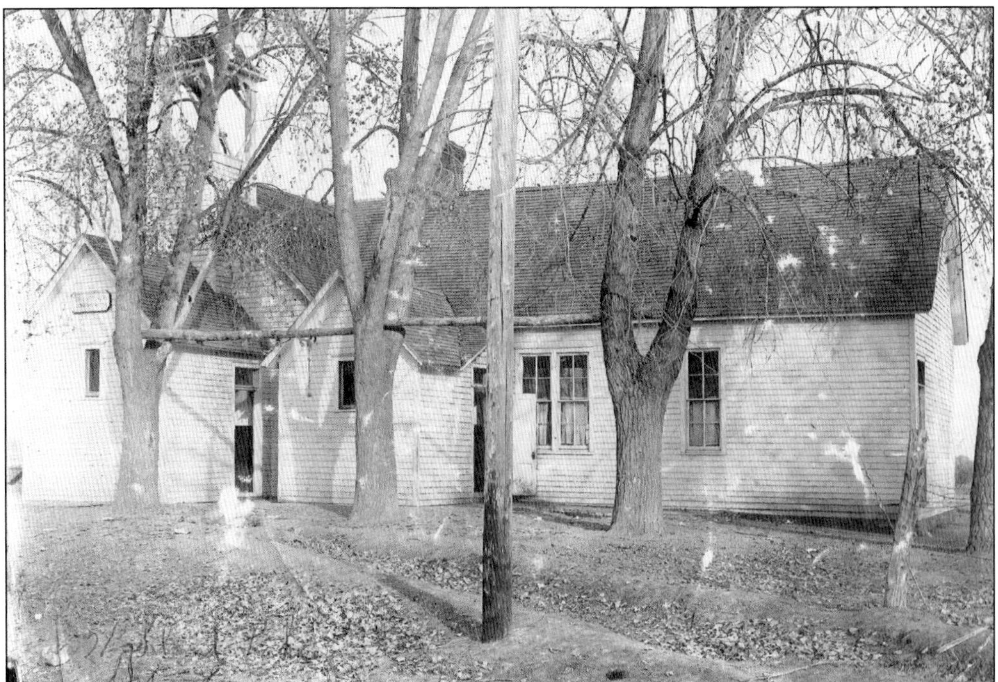

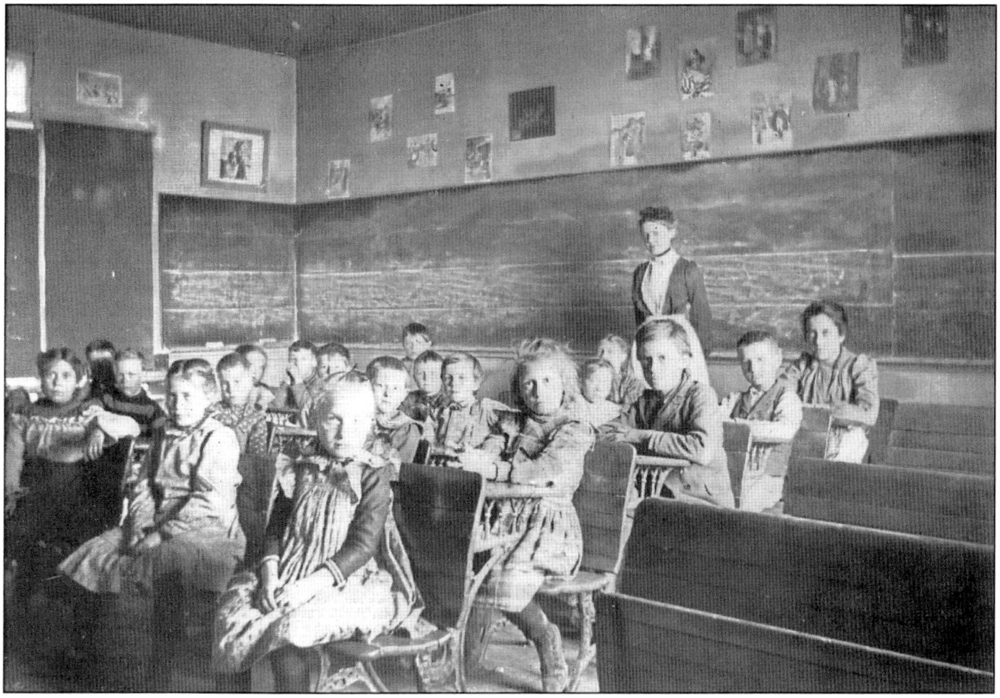

These photographs, taken in the 1890s, are the only known images of the interior of the Highlandlake School. The teacher is probably Elsa Strawbridge. The students are unidentified. The blackboards are made of panels of either pine or fir lumber and painted with a mixture of egg white and carbon from charred potatoes, a process first used in 1809. Teachers and students wrote with chunks of chalk and erased with cloth rags. The resultant chalk dust was the bane of all teachers. (Both, Mary Mead Jensen family collection, Historic Highlandlake.)

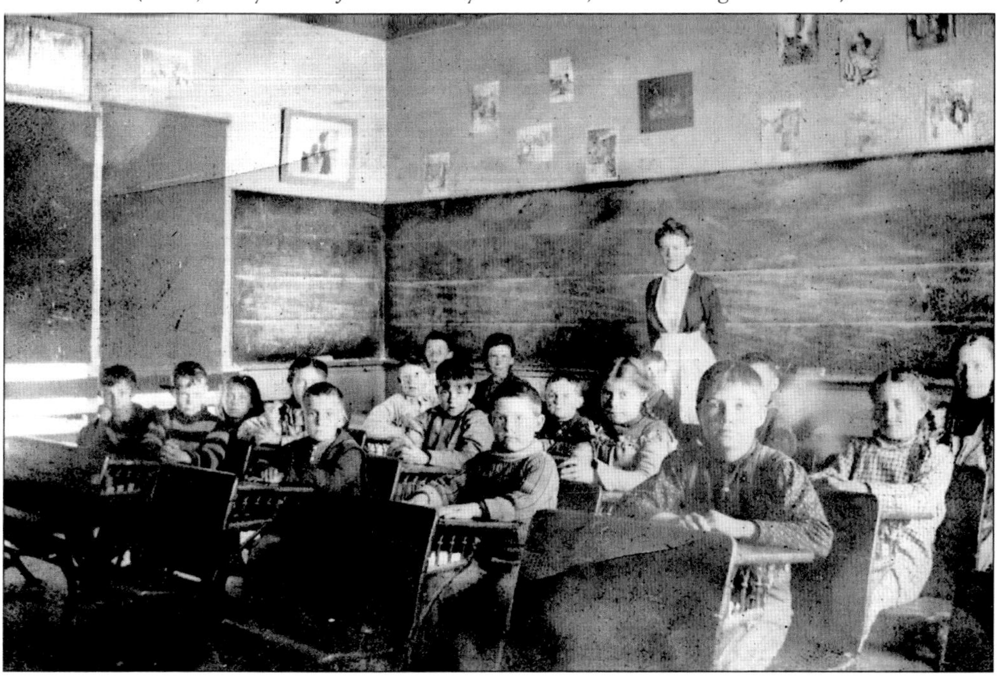

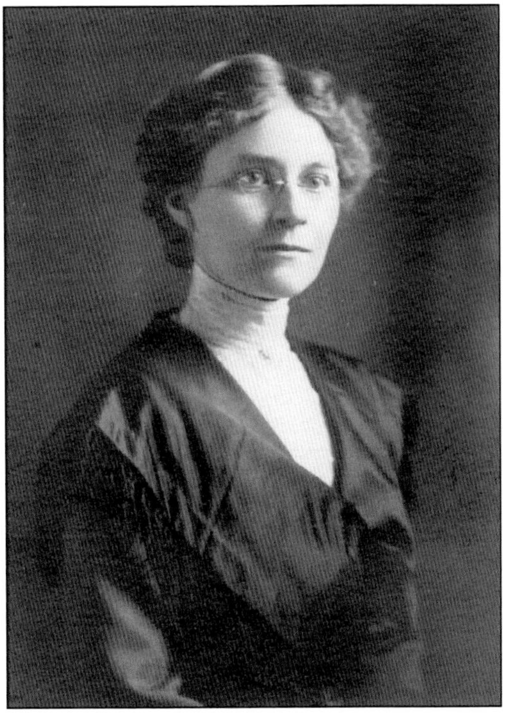

Elsa Strawbridge taught at the Highlandlake School in the late 1890s and early 1900s. While in Highlandlake, she met and married Rufus Mead, the youngest son of Lorin C. and Elizabeth Sheldon Mead. They later moved to Pasadena, California where Rufus became the superintendent of schools. (Mary Mead Jensen family collection, Historic Highlandlake.)

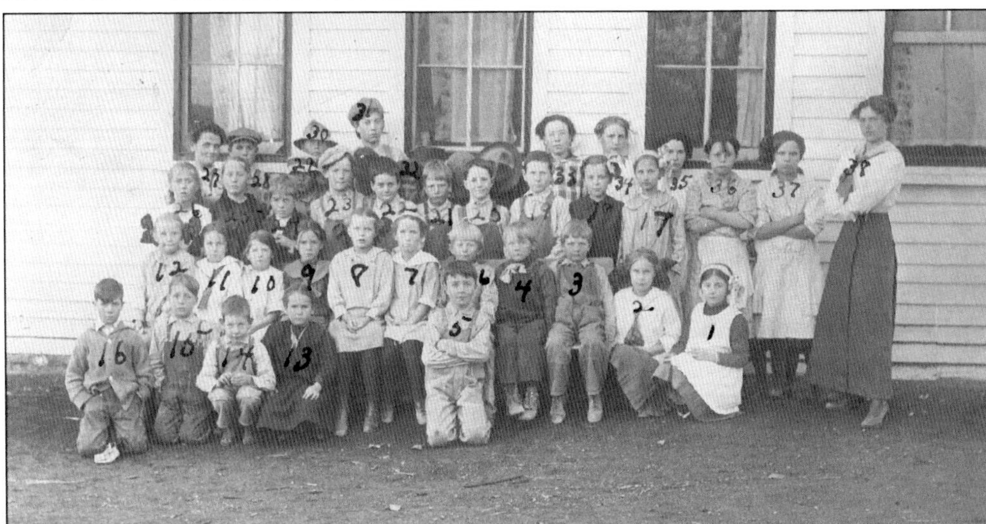

Shown are Mrs. Fuller and Blossom Entwhistle's Highlandlake classes from the 1915–1916 school year. Those pictured are numbered as follows: 1. Bessie Graham, 2. Ethel Markham, 3. Edward Brossman, 4. Estes Faulkner, 5. Ralph Kraig, 6. Roy Webber, 7. Roma Jepperson, 8. Mabel Bishop, 9. Georgene Pratt, 10. Phylis Emery, 11. Ruth Markham, 12. Ethelyn Bishop, 13. unidentified, 14. Orlando Hitt, 15. Albert Mudd, 16. Buddy Richard, 17. Marjorie Graham, 18. Dorothy Mead, 19. Charles Hitt, 20. Clyde True, 21. Elmer Brossman, 22. Roy Jones, 23. Alfred Martindale, 24. Anthony E. ?, 25. Ruby Faulkner, 26. Alvina Webber, 27. Mrs. Fuller, 28. Kenneth Graham, 29. Harold Mudd, 30. Frank Wokersin, 31. Glen Graham, 32. Charles Brossman, 33. Alice Kraig, 34. Florina Martindale, 35. Irma Pratt, 36. Ethel Pratt, 37. Opal Richard, and 38. Blossom Entwhistle. (Mary True-Farr collection, Historic Highlandlake.)

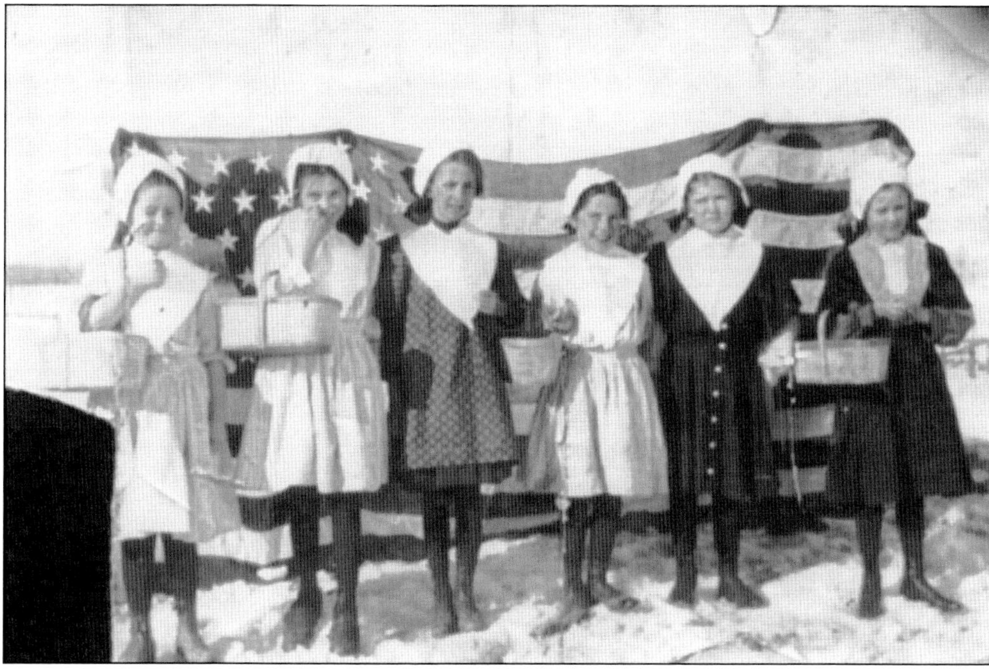

School plays were always popular community events. In the above photograph, six girls in Pilgrim costumes proudly pose in front of an American flag. The location was Highlandlake, and the year was 1915. Shown, from left to right, are Ethel Markham, Bessie Graham, two unidentified girls, Mabel Bishop, and unidentified. Shown below is the rest of the play's cast of characters, including Irma Pratt (left), Fiona Martindale (second from left), and Dorothy Mead (third from left); the other children are unidentified. (Both, Mary Mead Jensen family collection, Historic Highlandlake.)

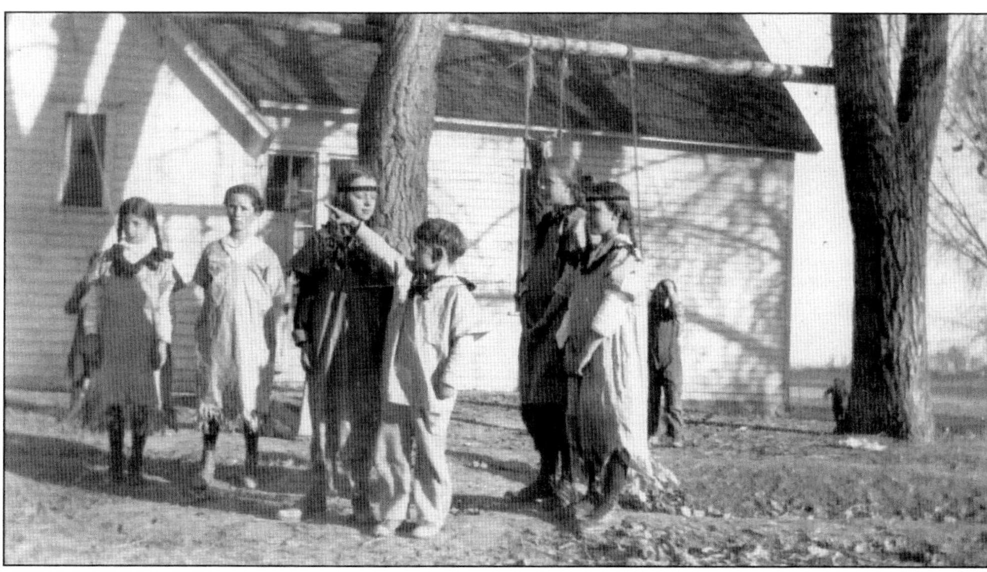

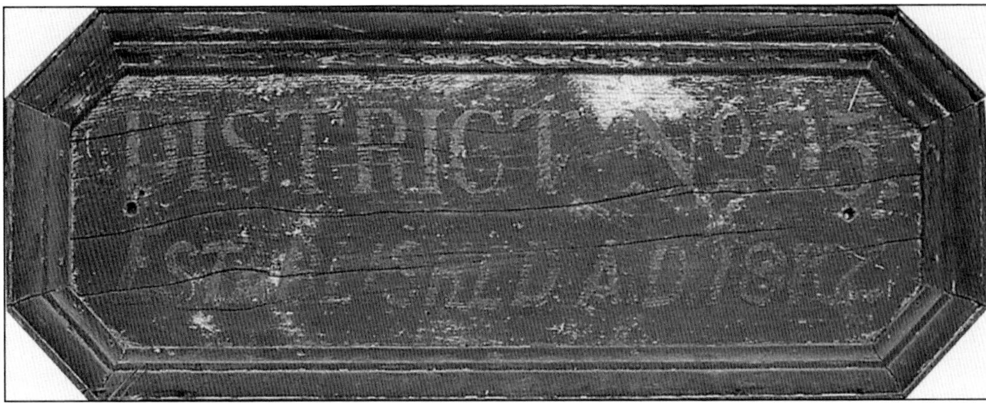

This sign from the Highlandlake School reads, "District No. 95, Established A.D. 1910." When Highlandlake District No. 33 merged into the new Mead Consolidated Schools District in 1910, the number changed. If one looks closely at the photograph on page 70, this sign hanging up in the front gable is visible. Shown below is the original Highlandlake School as it looks today. Once the Mead School District closed the school, the Edward Brossman family rescued the original one-room section and moved it to their property a block north of the school and used it first as a storage shed and later a garage. In 1995, Steve and Cheryl Senesac, unaware of the building's history, converted it into a preschool. Today, children come from all over the Mead area to attend this preschool. (Both, Pauli Driver Smith.)

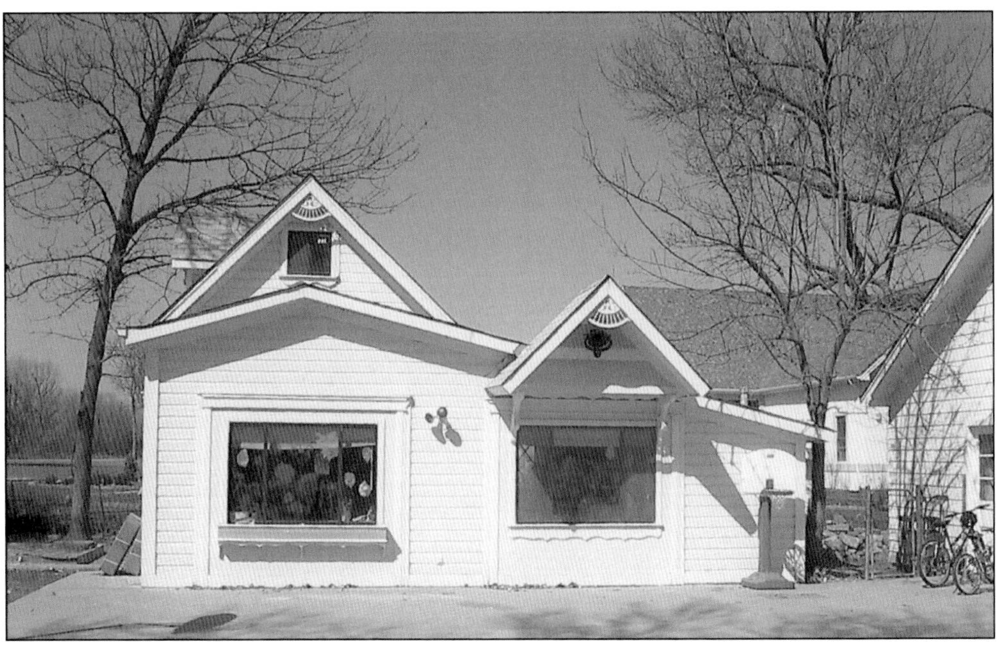

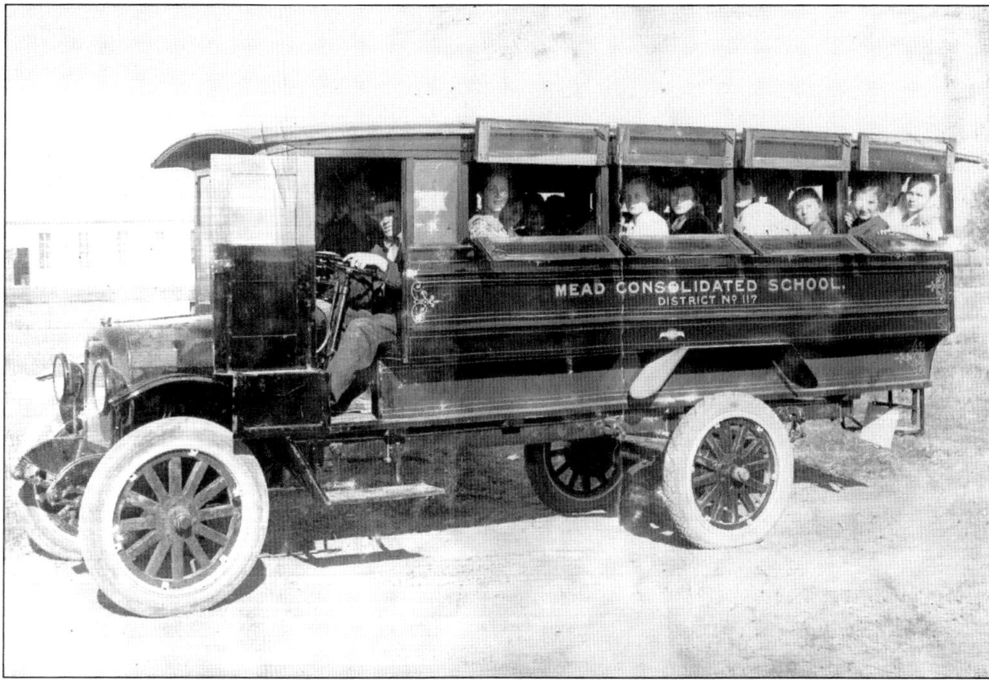

The Mead Consolidated Schools District covered a large geographical area. Children in town walked to school, and many in the outlying areas rode a horse or mule. Eventually, school buses became the preferred mode of transportation. This c. 1917 Model T Ford was the first bus used in the Mead Consolidated Schools District. Note the wood-framed windows. In the background, one can see the original two-room Mead schoolhouse. The bus driver, Marion Humphrey, is shown with his bus packed full of students. Below is a rare c. 1918 photograph of the new Mead Consolidated School building with three school buses parked in front. (Above, Hester Markham collection, Historic Highlandlake; below, Mary Mead Jensen family collection, Historic Highlandlake.)

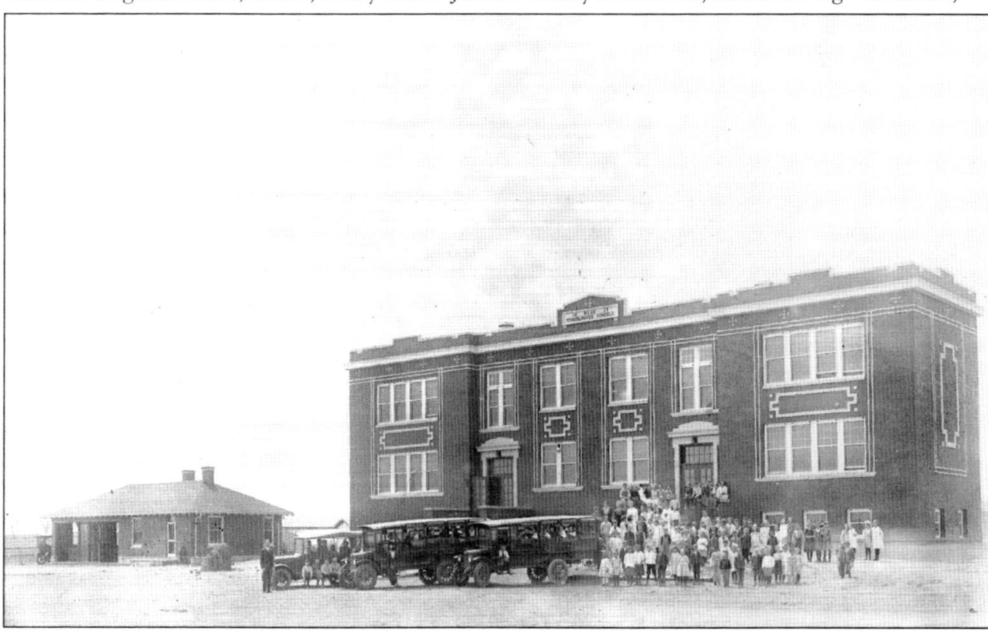

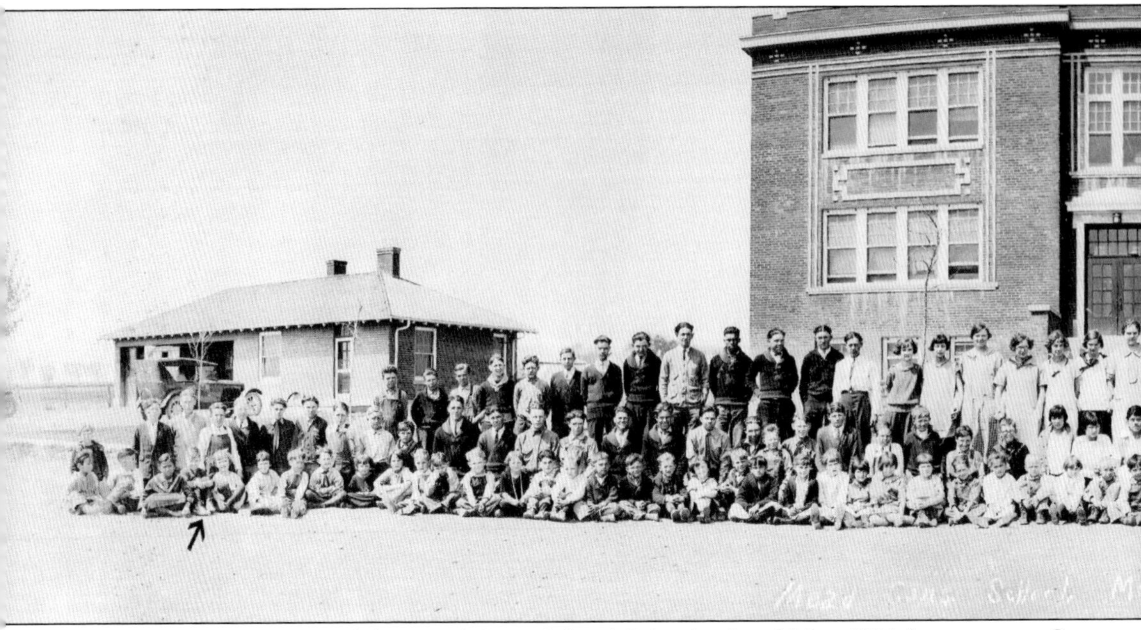

This panoramic view of the Mead Consolidated Schools, dated May 1925, has a secret. One enterprising young man appears twice. How did he do it? The photographer probably used a Circuit camera, a large boxlike machine mounted on a sturdy tripod, featuring a clocklike mechanism that would transport the film in one direction while the camera panned in the opposite direction. This exposed the film (quite slowly) as it traveled past the lens. According to

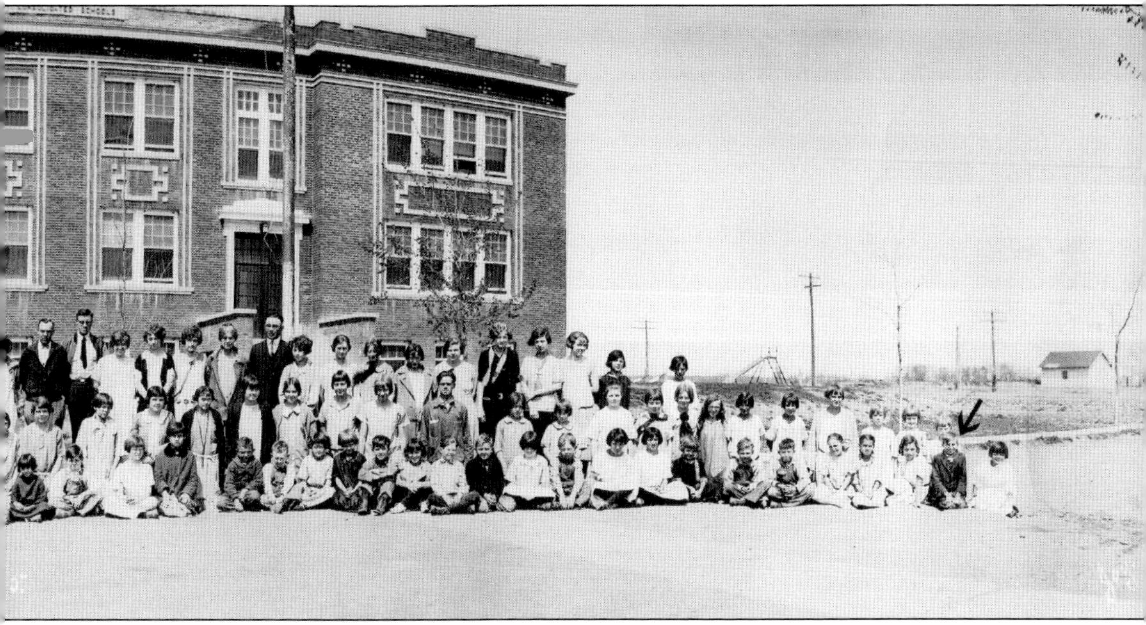

Glen Keller, who was present at the time, the boy stood near the far left of the group of children just long enough for the camera to pan past and then, while ducking behind his classmates, ran to the far right, popping up just in time for the camera to capture him again. (Glen Keller collection, Historic Highlandlake.)

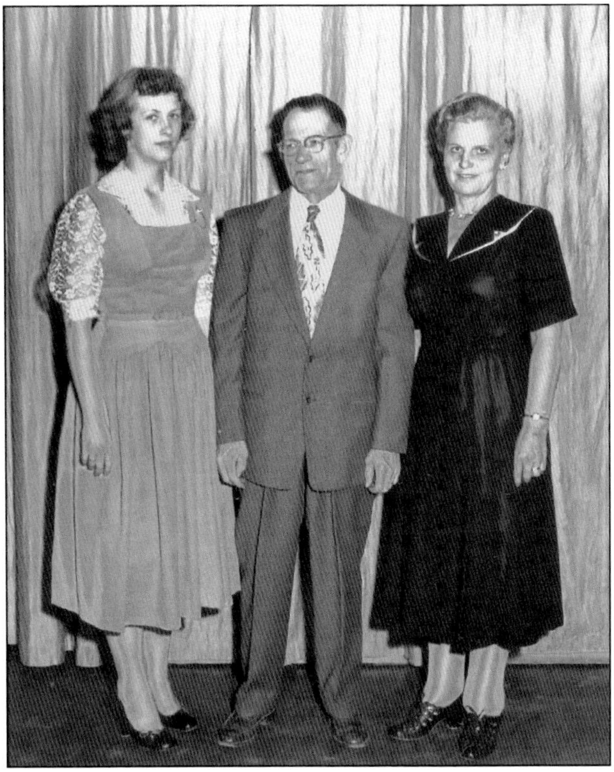

Marion and Esther Humphrey were the beloved school custodians. For decades, their job was to keep the school clean, make repairs as needed, and if that was not enough, Marion and occasionally Esther, drove a school bus. As their children grew, they also were enlisted to help with the work. Shown here, from left to right, are daughter Ethel and parents Marion and Esther Humphrey. (Historic Highlandlake.)

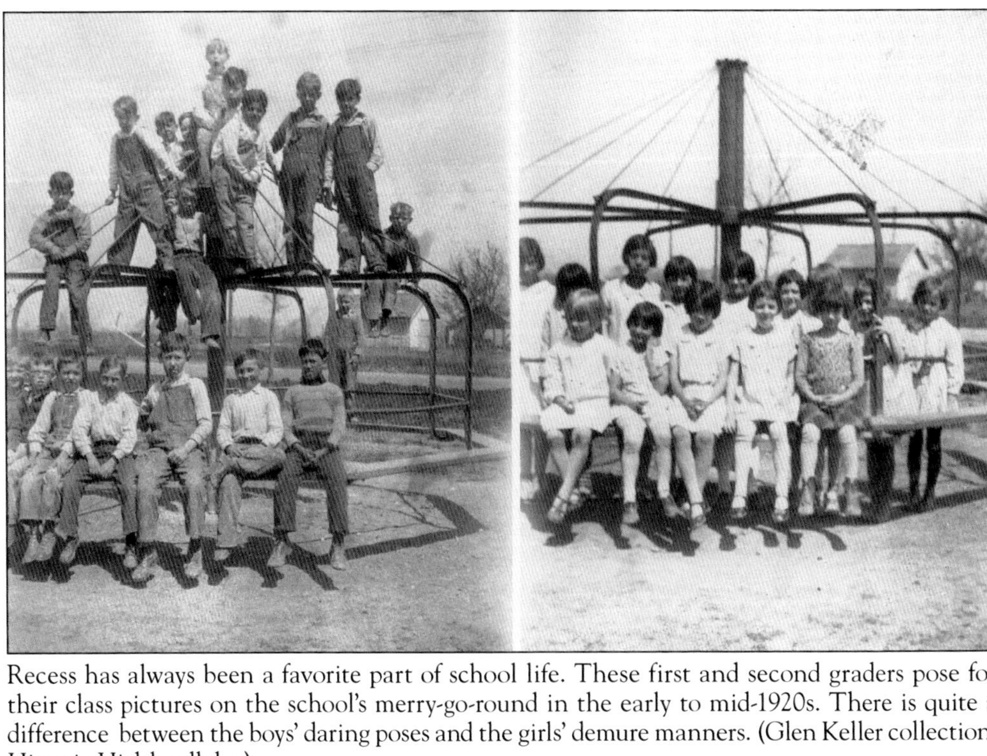

Recess has always been a favorite part of school life. These first and second graders pose for their class pictures on the school's merry-go-round in the early to mid-1920s. There is quite a difference between the boys' daring poses and the girls' demure manners. (Glen Keller collection, Historic Highlandlake.)

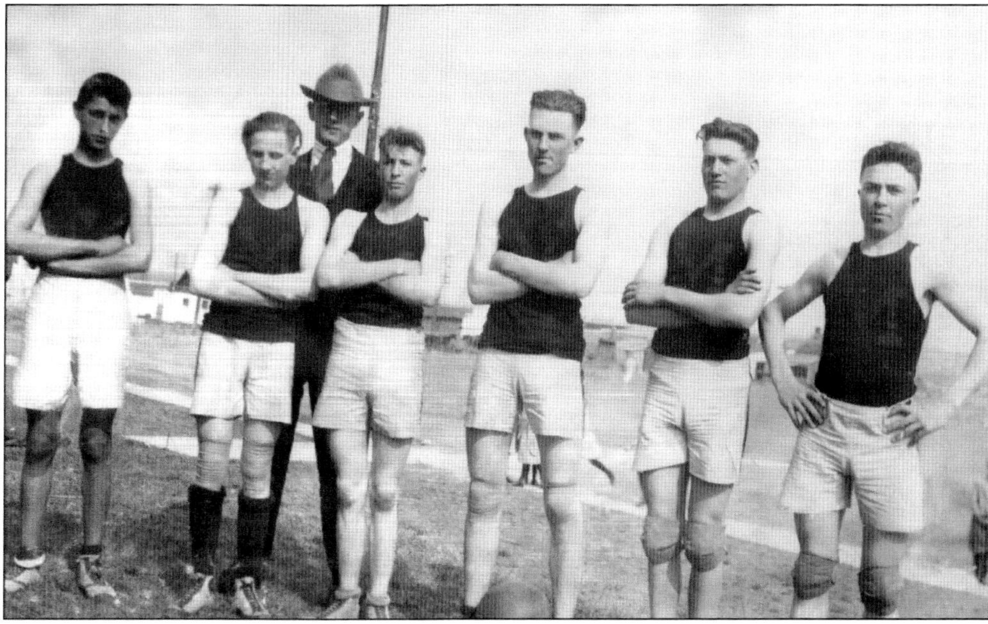

Being able to participate in organized sports was one of the benefits of the consolidated schools system, and both boys and girls took advantage of the opportunities. In this 1921 photograph, the Mead High School boys' basketball team poses with its coach. Harley Markham is third from the left, and the others are unidentified. (Hester Markham collection, Historic Highlandlake.)

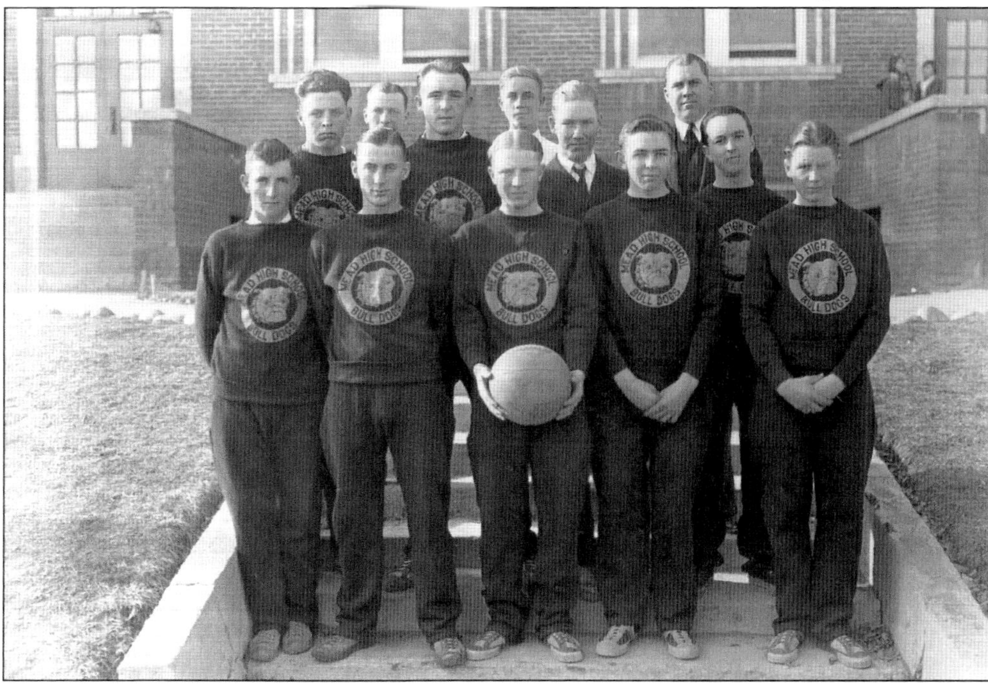

By the 1930–1931 school year, Mead basketball players sported the Mead High School Bulldog mascot on their sweat suits and uniforms. Note how their official portrait is also more formal than the one taken just 10 years before. In the first row, fourth from the left, is Keith Coates. Ralph Hanson is in the back row, fifth from the left. (Loren Coates collection, Historic Highlandlake.)

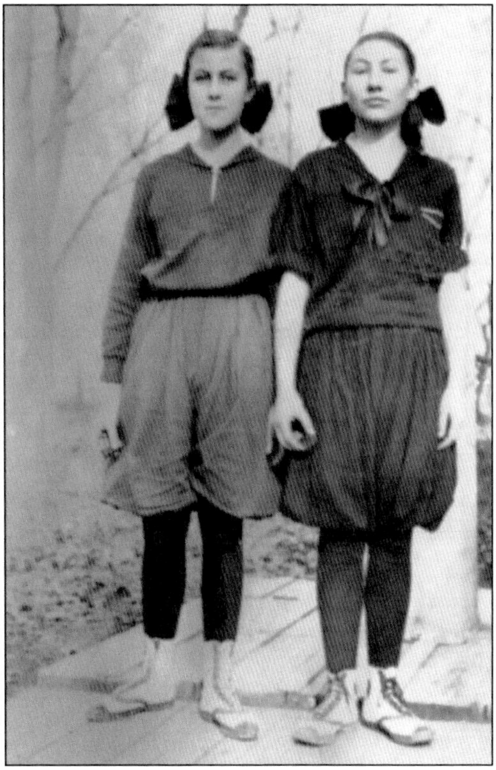

Starting in 1920, the Mead school allowed girls to play basketball. Their mismatched uniforms consisted of a middy blouse and bloomers sewn from several yards of fabric. Black cotton leggings and high-top sneakers completed the outfit. Shown on the left is eighth-grader Ruth Markham, who played running center, and on the right is ninth-grader Dorothy Pearcy, who played forward. Their coach (not shown) was school superintendent B.V. McCready. Shown below is freshman Ruth Markham in her 1921 uniform, which includes the school's name running diagonally across the front of her blouse. Her leggings have been traded in for knee socks. (Both, Jean Howe collection, Historic Highlandlake.)

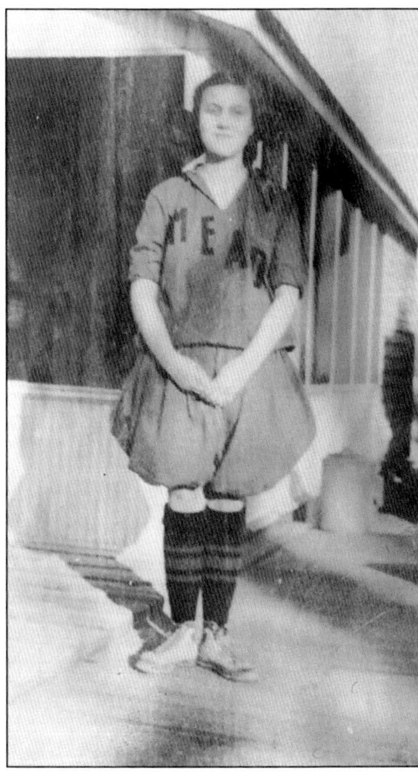

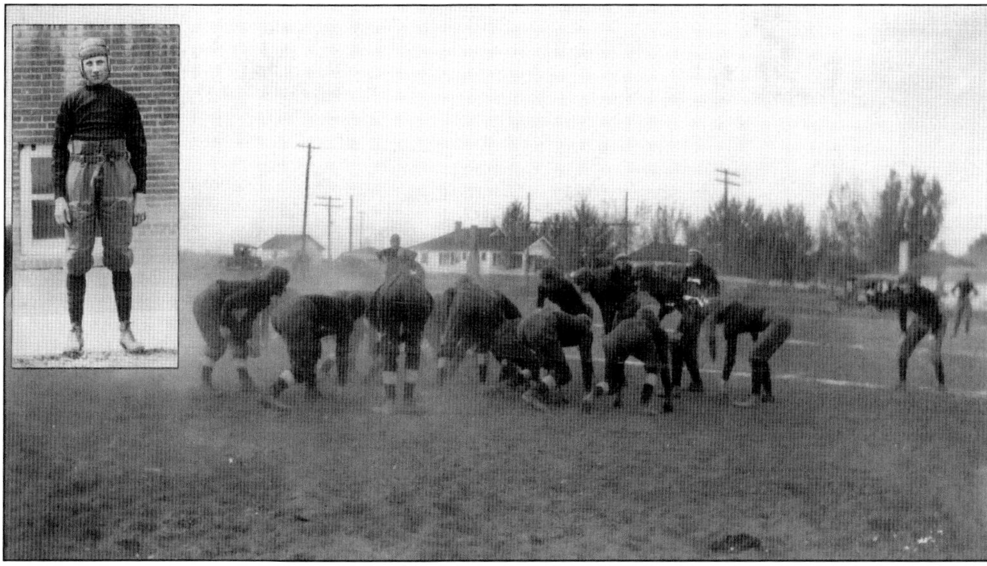

Football was as popular a school sport in 1922 as it is today. However, there is a huge difference in the uniforms worn in the 1920s and those worn by today's students. Charles Hitt proudly wears his "leatherhead" helmet and high-waisted pants, and his teammates line up for the hike in the larger image. Below, the team scrambles for the football. Protective gear, not to mention playing fields, hardly resembled today's heavily padded uniforms and manicured lawns. Mead's field was a dirt patch filled with goat head stickers and thorns. Woe to the hapless player who fell, for the thorns always managed to find their way into a hand or thinly protected knee. (Both, Jean Howe collection, Historic Highlandlake.)

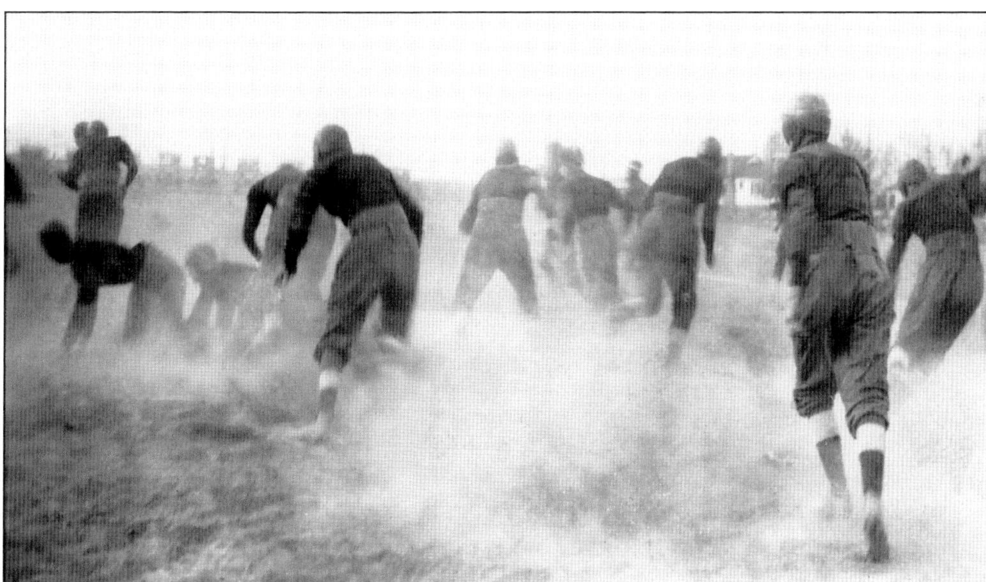

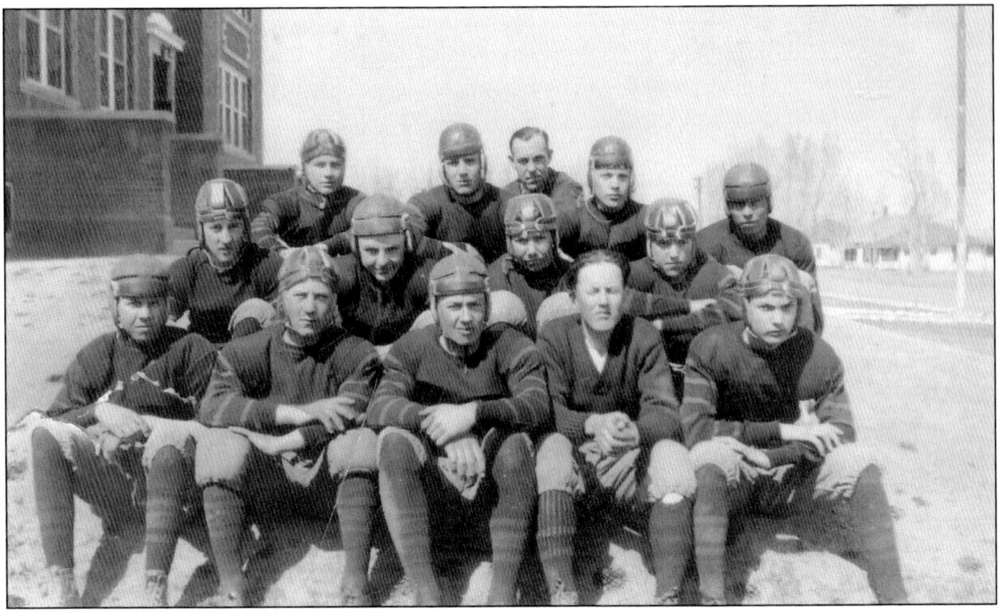

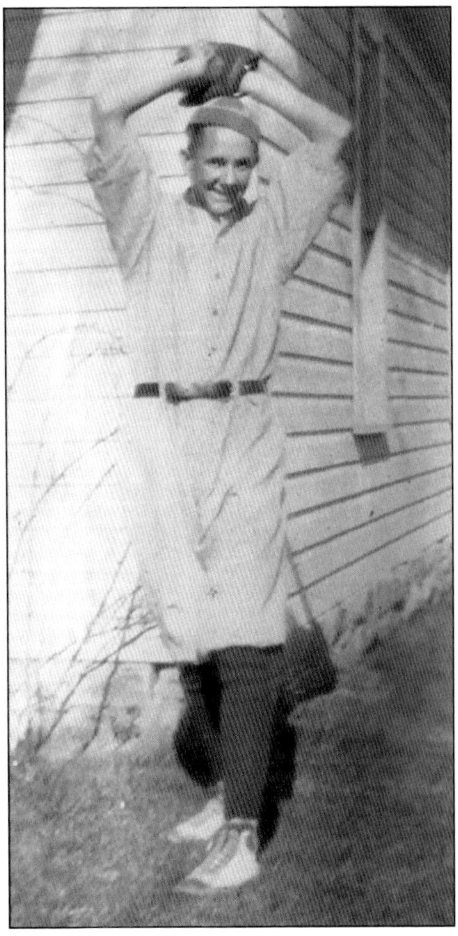

As Mead Consolidated School entered its second decade, there were few noticeable changes to the high school's football uniforms. Students were still wearing sweaters, minimal padding, and no numbers. The only notable differences seen are that some of the men are now wearing a newer style of helmet. Wayne Coates is in the second row, second from the left. (Loren Coates collection, Historic Highlandlake.)

Baseball is the third popular men's sports played at the Mead School. Sometime in the spring of 1923, Ray S. Tyson struck an exaggerated pitcher's pose for Ethel Markham and her camera. This is one of only two known photographs of students in their baseball uniforms from this era. Markham commemorates the event in her 1923 graduation journal. (Jean Howe collection, Historic Highlandlake.)

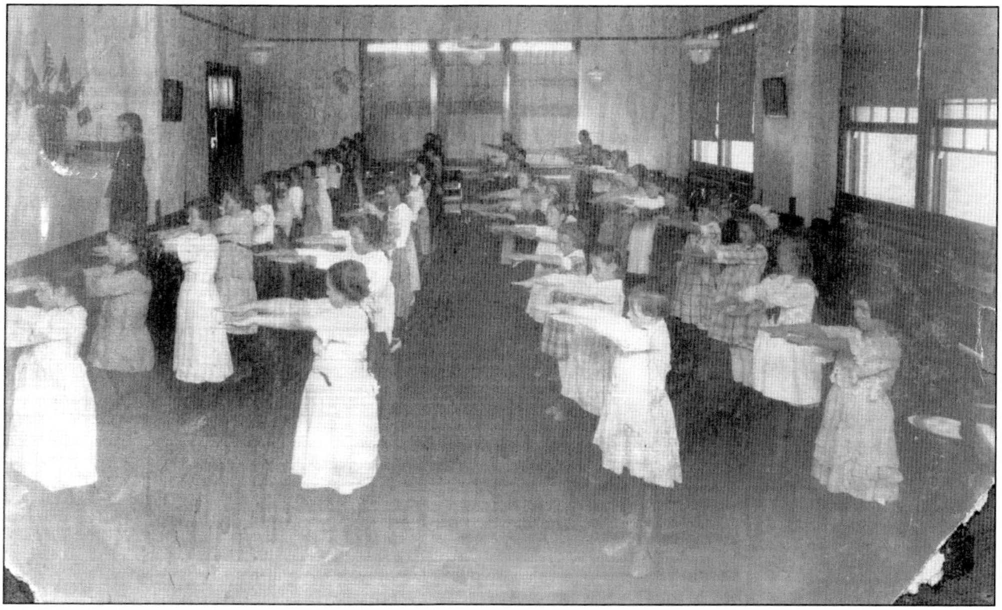

Physical education classes were a part of the curriculum. Here, young girls go through their paces in the Farmer's Union Hall, which doubled as a gym for the school until the Works Progress Administration (WPA), in the mid-1930s, built a gym at the school. Prior to that, children walked a block north on Main Street to attend their physical education classes. (Cotton Johnson collection, Historic Highlandlake.)

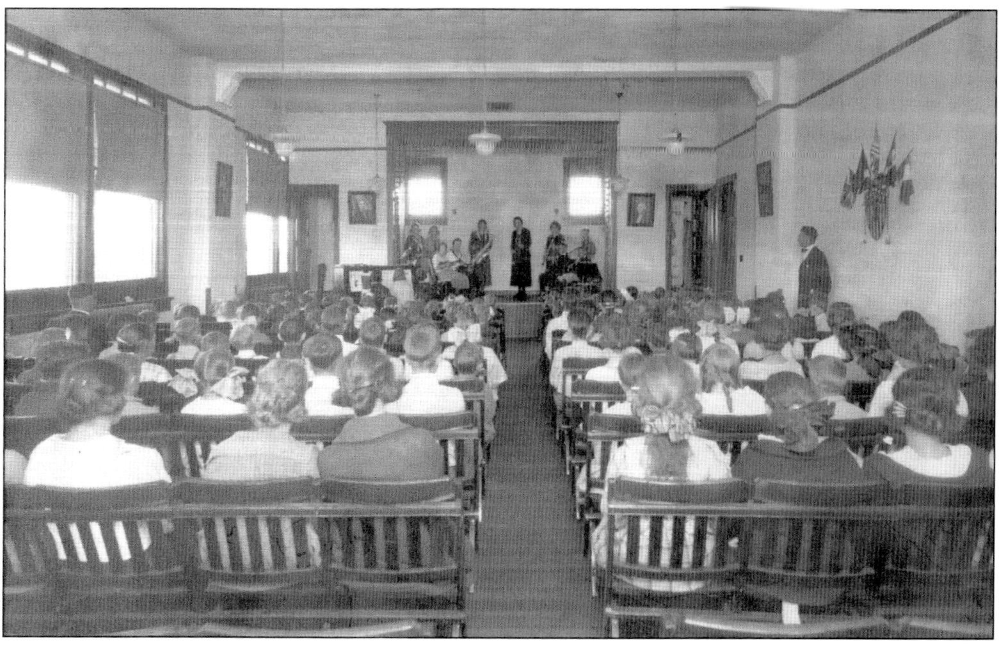

The Farmer's Union Hall, located a block north of the school on Main Street, hosted several events throughout the school year. Besides gym classes, there were also school assemblies, plays, and concerts. In this photograph, several students are playing instruments for what appears to be a school assembly, while school superintendent B.V. McCready watches from the sidelines. (Cotton Johnson collection, Historic Highlandlake.)

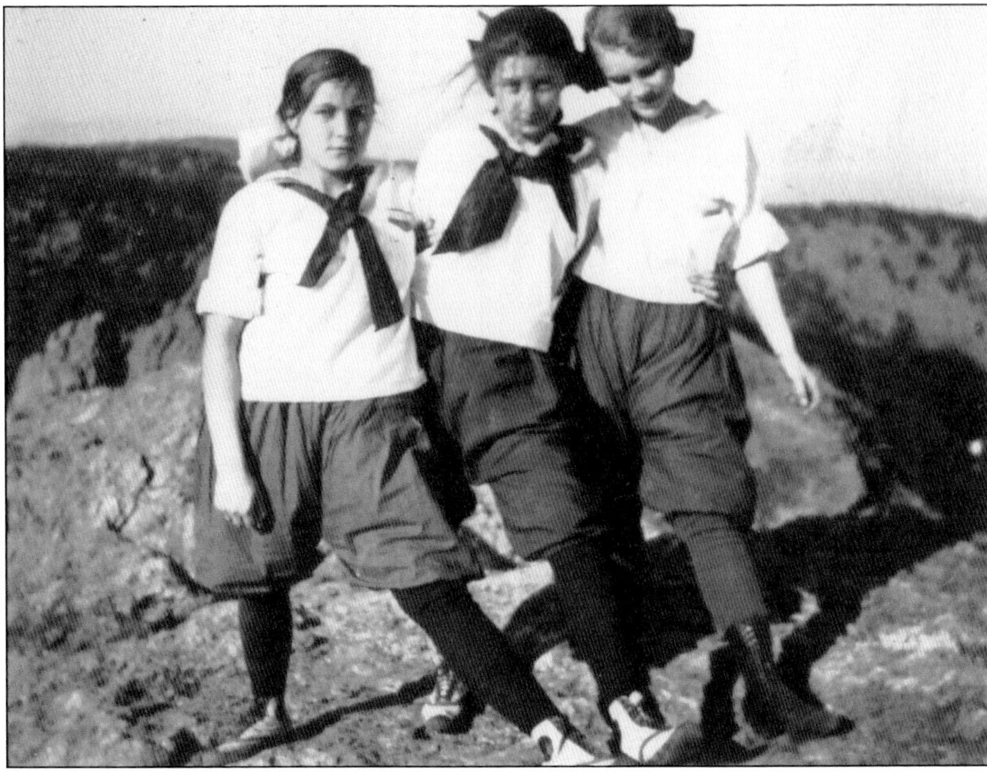

Three Camp Fire Girls pose during an outing to the nearby mountains in 1922. After finals week, hiking brought a welcome relief from the stresses of studying. Camp Fire Girls of America was founded in 1910 as a sister organization to the Boy Scouts of America. It was the first nonsectarian, multicultural organization for girls in America. The programs emphasized camping and other outdoor activities. Dorothy Mead is on the far right. The other two girls in the photograph are not identified. (Mary Mead Jensen family collection, Historic Highlandlake.)

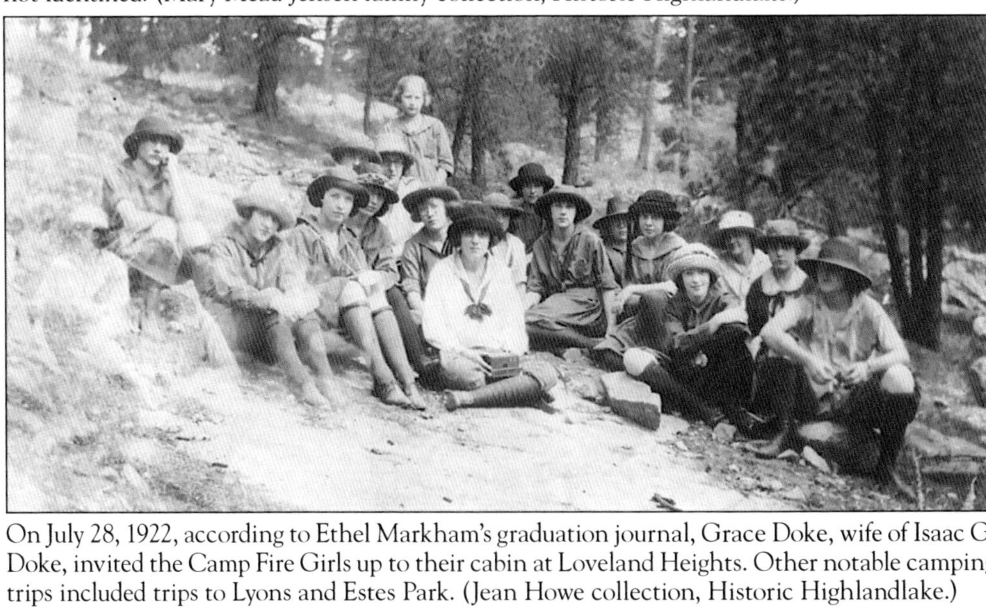

On July 28, 1922, according to Ethel Markham's graduation journal, Grace Doke, wife of Isaac G. Doke, invited the Camp Fire Girls up to their cabin at Loveland Heights. Other notable camping trips included trips to Lyons and Estes Park. (Jean Howe collection, Historic Highlandlake.)

Published in October 1921, issue No. 9 of *The Mead School "Pep"* is a six-page call to adjoining districts to consolidate into the Mead District. Some of the advantages listed for joining included the following: "More individual attention to each child, better-trained teachers, directed plays and games, and classes in music, drawing, manual training, household arts, and agriculture." (Mary Mead Jensen family collection, Historic Highlandlake.)

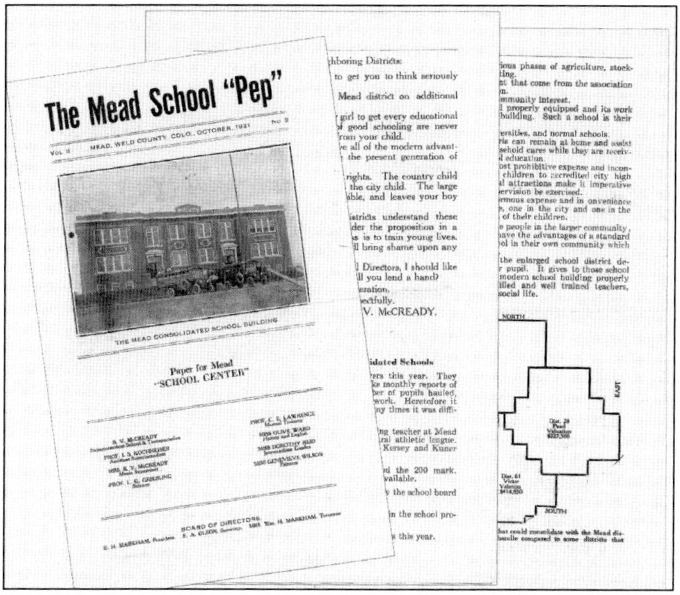

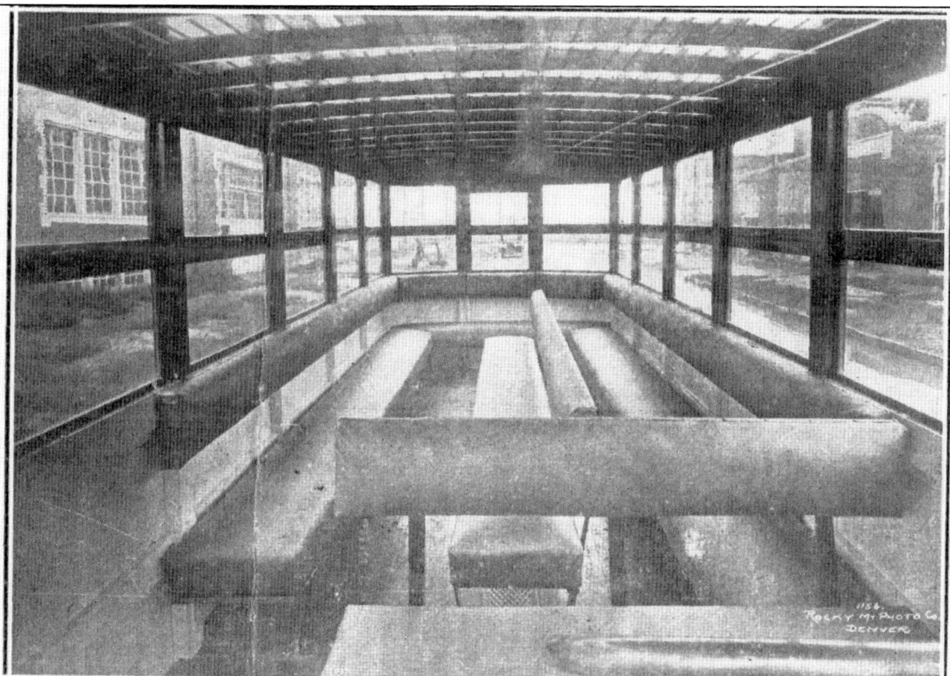

Above is a cut of interior of the Mead School Bus No. 2. 45 can be seated comfortably. Last month's report showed it cost 11 cents per day for each child transported. This included depreciation on truck, wear on tires, gas and oil consumption, storage, driver and all expenses. If you live along a trolley line and your children rode for 11 cents per day you would say it was a minimum charge.

One page in the *Mead School "Pep"* includes an interior shot of Mead School bus No. 2. The caption states that the bus seats 45 students comfortably and as of the last month's report, the cost was 11¢ per day, per child. Costs included depreciation on the truck, tire wear, gas and oil consumption, storage, driver, and all expenses. What a deal! (Mary Mead Jensen family collection, Historic Highlandlake.)

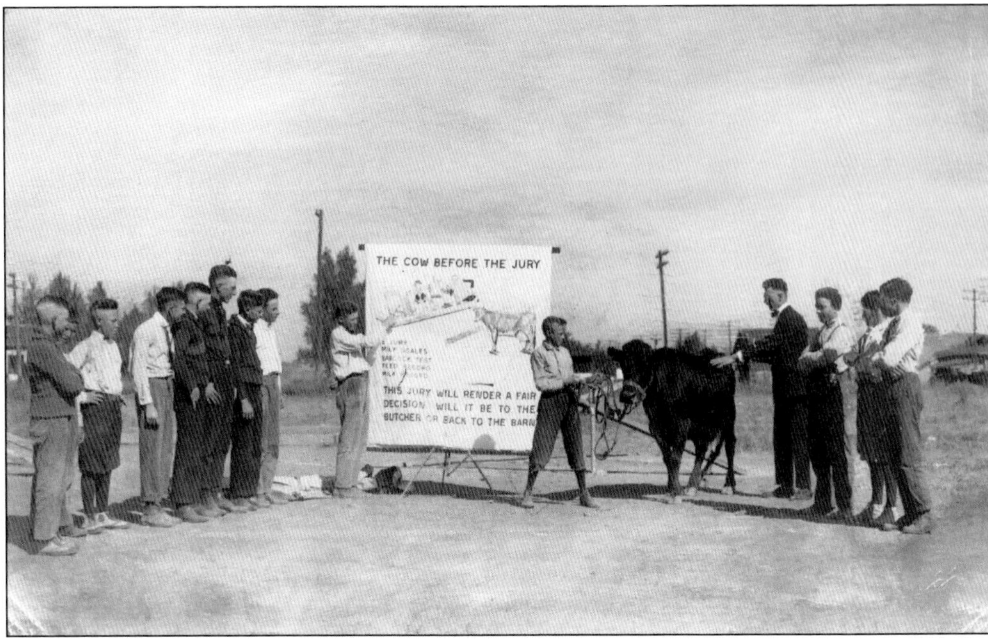

The Mead Consolidated School offered an assortment of classes aimed at preparing students for future careers. In this undated photograph, "The Cow before the Jury," school superintendent B.V. McCready helps students decide the fate of a young heifer. "This jury," the sign proclaims, "will render a fair decision. Will it be to the butcher or back to the barn?" (Cotton Johnson collection, Historic Highlandlake.)

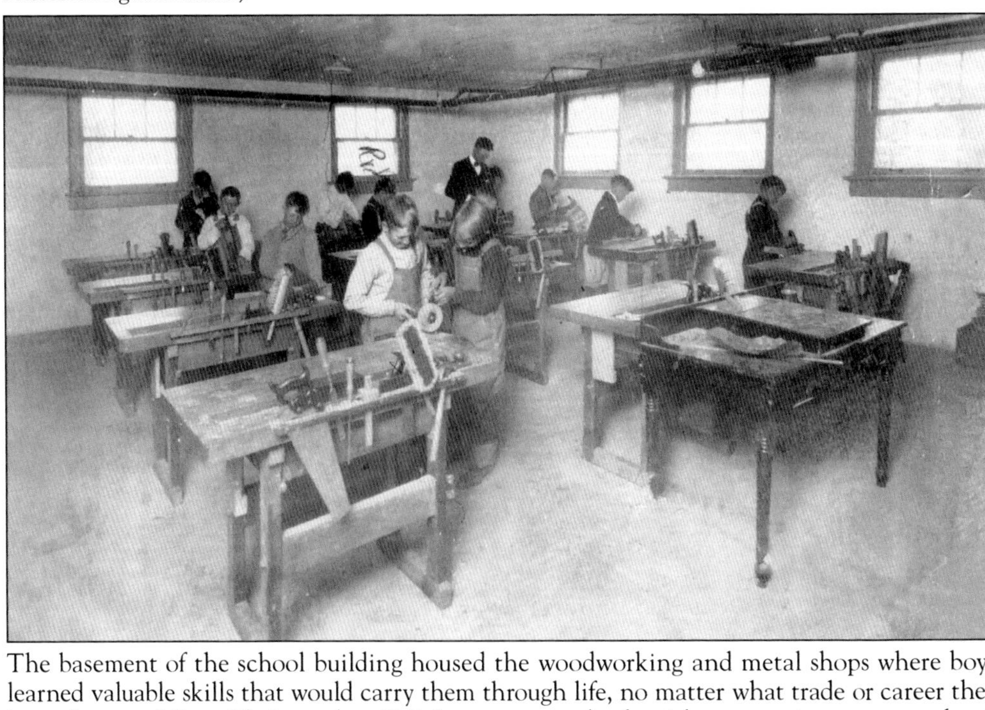

The basement of the school building housed the woodworking and metal shops where boys learned valuable skills that would carry them through life, no matter what trade or career they later chose to follow. Their teacher, C.E. Lawrence, in the far right corner, inspects a student's work for accuracy. (Cotton Johnson collection, Historic Highlandlake.)

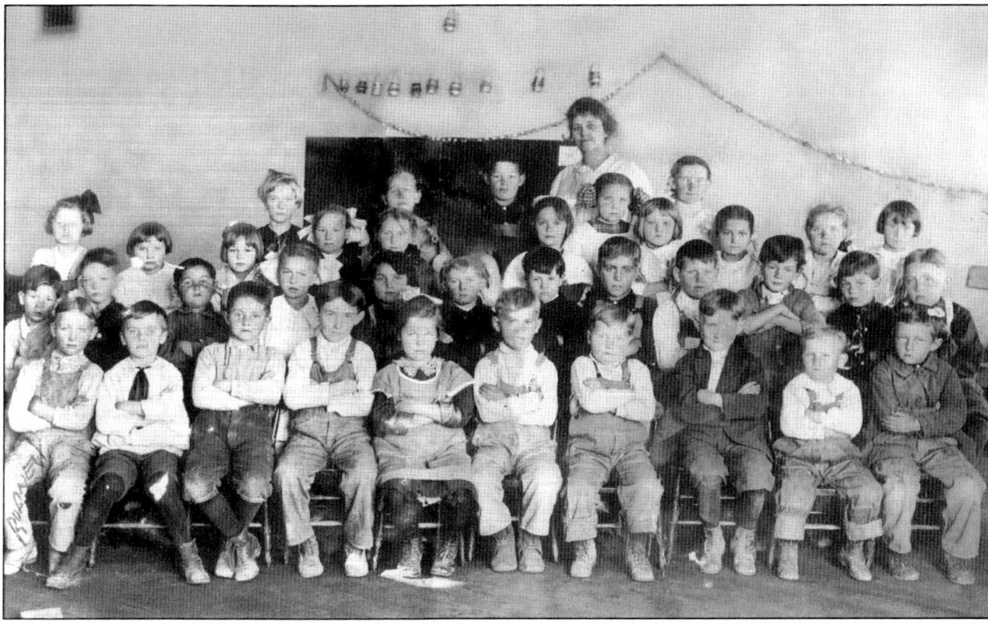

Tradition dictates that students dutifully pose each year for their official class photograph. For this 1916 class portrait, some students appear resigned to their fate while others mug for the camera. Rodney Markham, looking like a young Tom Sawyer, complete with a hole in the knee of his pants, is the first child on the left in the first row. The Mead School, built in 1906, is the most likely location. (Hester Markham collection, Historic Highlandlake.)

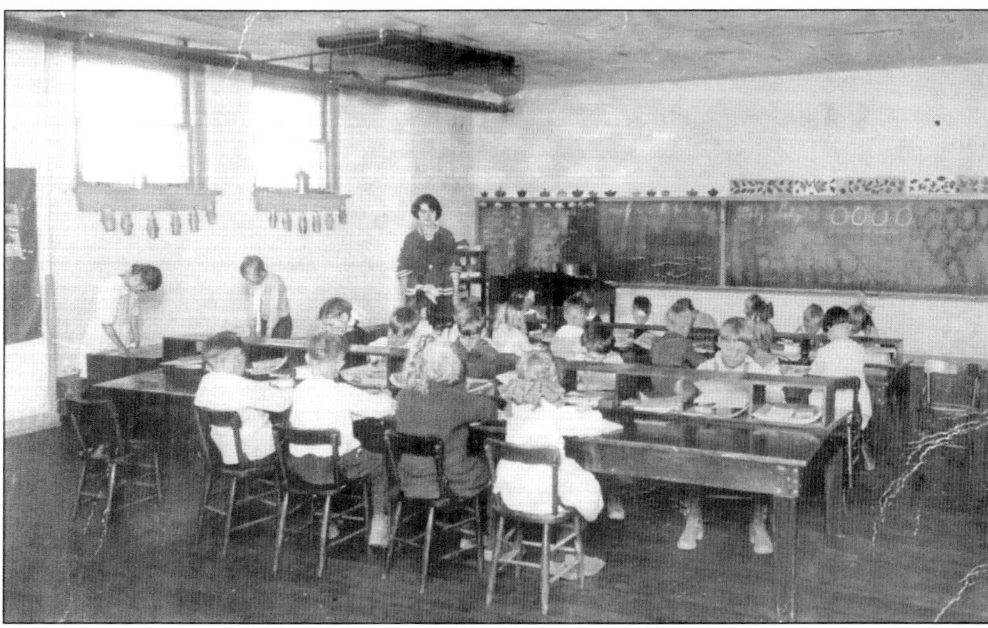

In the first- and second-grade classrooms, each desk had a shelf with cubicles running down the center. The children used these to place their school supplies and books. Hester Markham, who taught at the Mead Consolidated School for many years, said that the boys would tease the girls by grabbing their schools supplies from the cubicles. (Cotton Johnson collection, Historic Highlandlake.)

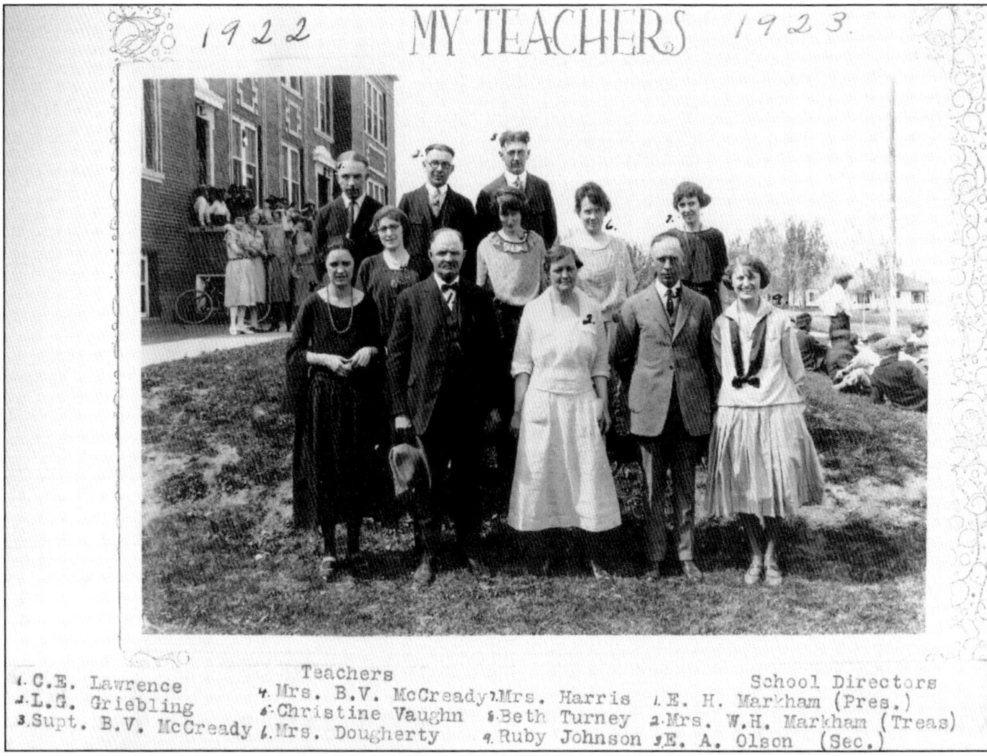

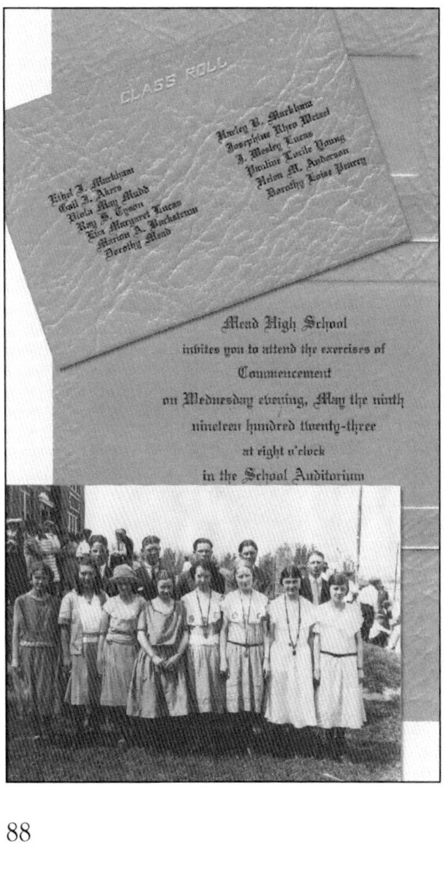

For a high school graduation present, Ethel Markham's parents presented her with a beautiful graduation journal. Markham filled every one of the journal's pages with memories of her senior year, including many photographs of herself, her classmates, school activities, and teachers. On this page, she pasted a photograph of her teachers and school directors, including her mother, Pearl Markham, and uncle Edmond Herbert Markham. (Jean Howe collection, Historic Highlandlake.)

The Mead Consolidated School senior class of 1923 sent out a formal announcement of its upcoming commencement exercises. Inserted inside was a complete list of the graduates. In the accompanying photograph, the proud seniors pose in front of the school for the camera. Other students and faculty stand around in the background, waiting for their turns. Pictured are, from left to right, (first row) Ethel Markham, Dorothy Louise Pearcy, Pauline Lucile Young, Eva Margaret Lucas, Viola May Mudd, Helen M. Anderson, Josephine Rhea Wetzel, and Dorothy Mead; (second row) J. Wesley Lucas, Gail Akers, Ray S. Tyson, Marion A. Backstrom, and Harley Markham. (Jean Howe collection, Historic Highlandlake.)

Five

Faith of our Fathers and Mothers

The two earliest churches in the area are the Congregational Church of Highlandlake, established in November 1880 by Lorin C. Mead, and the Mount Zion United Brethren Church, founded early the next year as a mission of the United Brethren Church. The Highlandlake congregation met in the Highlandlake schoolhouse until December 1896, after which it moved to its new church building on land donated by Lorin C. Mead. It is unclear where the Mount Zion congregation met until constructing a building on land donated by John Thompson of Longmont around 1883–1885, but it is assumed that they also met in a local schoolhouse, a common practice at the time.

In 1906, the Mount Zion members moved to Mead and renamed their church Mead United Brethren in Christ Community Church. They eventually sold the original church building to the Catholic Church and moved the two-story parsonage to the new location at the intersection of Fifth Street and Palmer Avenue. Proceeds from the sale of the old building helped to finance the construction of a new church. In 1946, the United Brethren and the Evangelical churches merged to form the Evangelical United Brethren Church (EUB). In 1968, the EUB church merged with the Methodist church to become the United Methodist Church. Once again, the church's name changed, this time to First United Methodist Church, Mead.

In late 1910, the Catholic church moved the old Mount Zion Church building to a lot in Mead donated by the Denver Elevator Company and renamed it Guardian Angels Catholic Church. In 1929, the length was extended to accommodate more seating, and a choir loft was added in the 1940s. A new parish hall was built in 1964 where Mass was held until the 1990s, after which the church building was restored. Both Mead churches still have active congregations. However, after 1917, the Highlandlake church only met sporadically until permanently closing in 1921. The fully restored Highlandlake Church building, now owned by Historic Highlandlake, Inc., is a community center and wedding chapel.

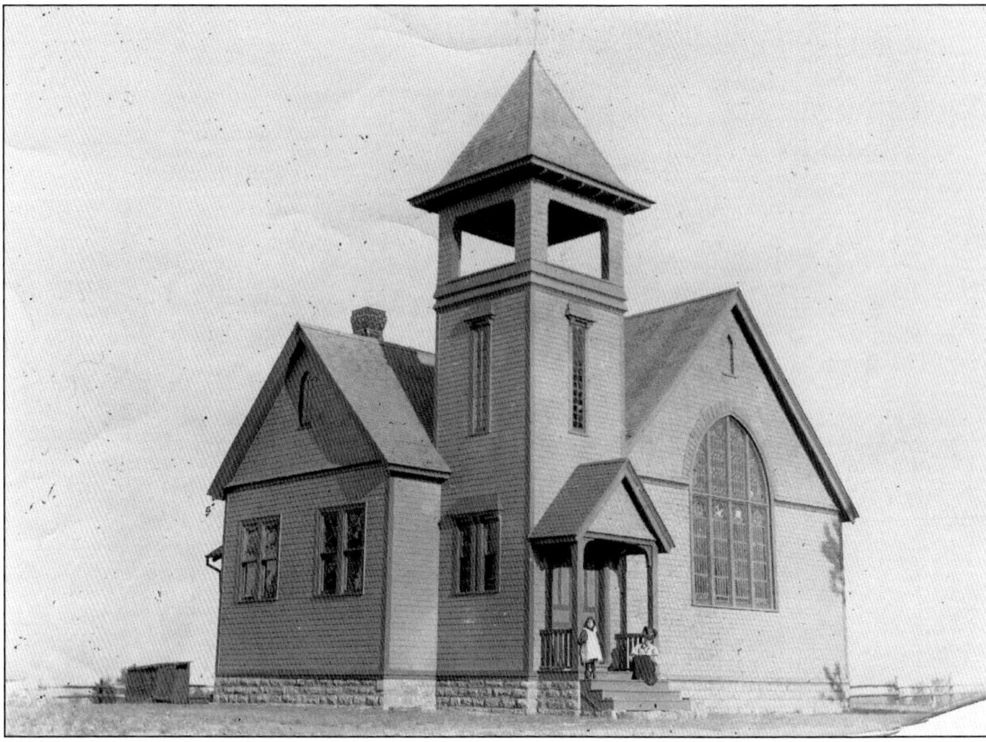

In the fall of 1895, Rev. Mary Bumstead started raising funds to build a church in Highlandlake. They laid the cornerstone on May 10, 1896, and finished the building in September. Between September and December, the congregation raised enough additional money to pay for wool carpeting, seating, lights, a potbelly stove, and an organ. Reverend Bumstead is sitting on the steps, but the child is unidentified. (Longmont Museum and Cultural Center.)

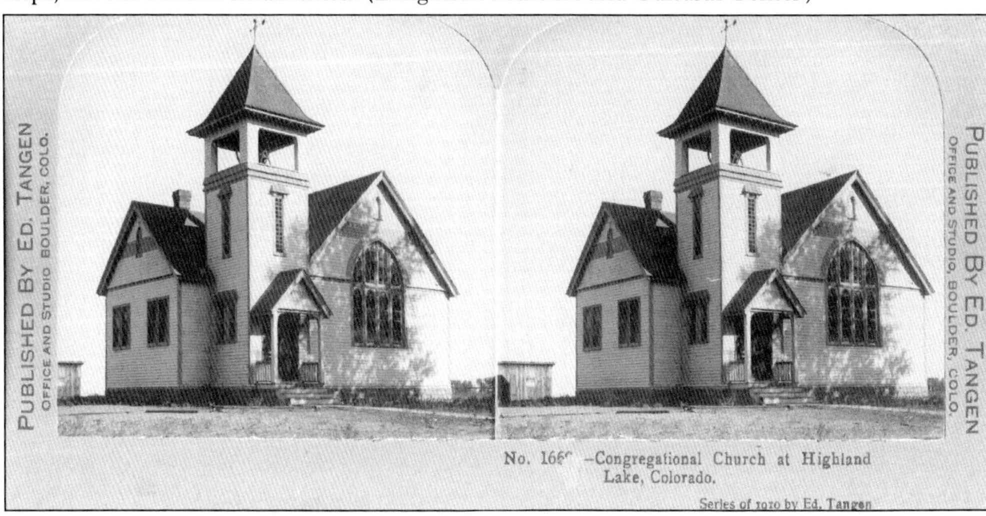

This stereographic card by Edward Tangen showing the Congregational Church of Highlandlake was published in 1910. This same photograph appeared in the September 6, 1906, special edition of the *Longmont Ledger*. Between 1896 and 1921, the church was painted four different ways. The photograph clearly shows the second paint scheme. Note the stripes in the gables. (Mark French collection, Historic Highlandlake.)

Rev. Mary Bumstead came to Highlandlake the summer of 1894. Until her sudden death in March 1898, she oversaw the building of the Highlandlake Church, served as president of the Missionary Society, occasionally taught school when one of the teachers was ill, and, in a pinch, served as the community's midwife. (Mary Mead Jensen family collection, Historic Highlandlake.)

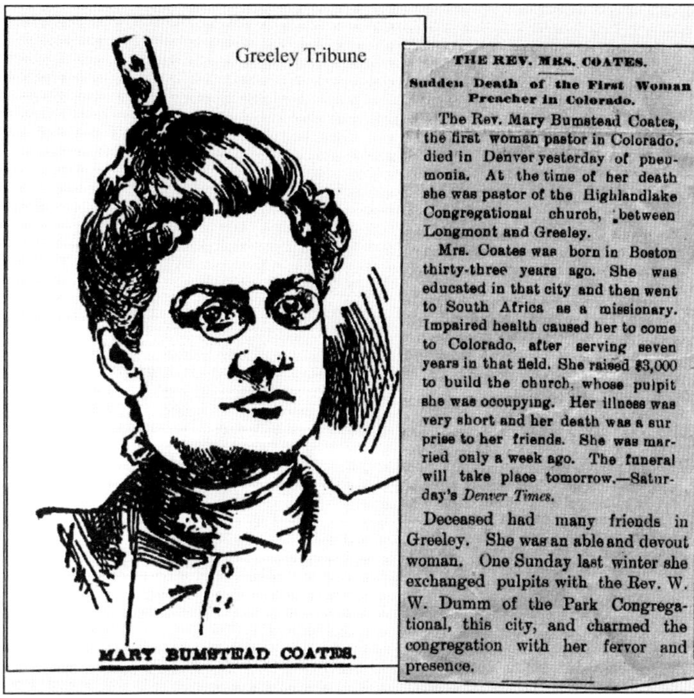

THE REV. MRS. COATES.

Sudden Death of the First Woman Preacher in Colorado.

The Rev. Mary Bumstead Coates, the first woman pastor in Colorado, died in Denver yesterday of pneumonia. At the time of her death she was pastor of the Highlandlake Congregational church, between Longmont and Greeley.

Mrs. Coates was born in Boston thirty-three years ago. She was educated in that city and then went to South Africa as a missionary. Impaired health caused her to come to Colorado, after serving seven years in that field. She raised $3,000 to build the church, whose pulpit she was occupying. Her illness was very short and her death was a surprise to her friends. She was married only a week ago. The funeral will take place tomorrow.—Saturday's *Denver Times*.

Deceased had many friends in Greeley. She was an able and devout woman. One Sunday last winter she exchanged pulpits with the Rev. W. W. Dumm of the Park Congregational, this city, and charmed the congregation with her fervor and presence.

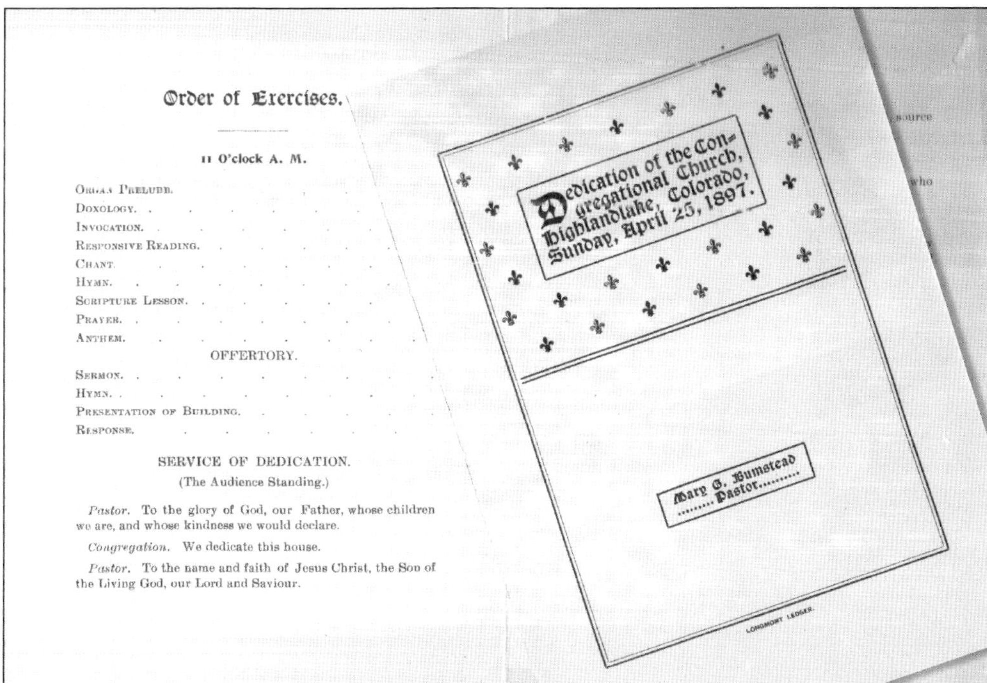

Order of Exercises.

11 O'clock A. M.

Organ Prelude.
Doxology.
Invocation.
Responsive Reading.
Chant.
Hymn.
Scripture Lesson.
Prayer.
Anthem.

OFFERTORY.

Sermon.
Hymn.
Presentation of Building.
Response.

SERVICE OF DEDICATION.

(The Audience Standing.)

Pastor. To the glory of God, our Father, whose children we are, and whose kindness we would declare.

Congregation. We dedicate this house.

Pastor. To the name and faith of Jesus Christ, the Son of the Living God, our Lord and Saviour.

The April 25, 1897, dedication of the Congregational Church of Highlandlake occurred nearly one year from when the cornerstone was first laid. Members of the congregation were happily surprised when Reverend Bumstead announced that not only were the building and its furnishings paid in full, but that Mrs. Rev. Charles H. (Augusta) Tainter of Chicago had paid off the mortgage on the church parsonage. (Historic Highlandlake.)

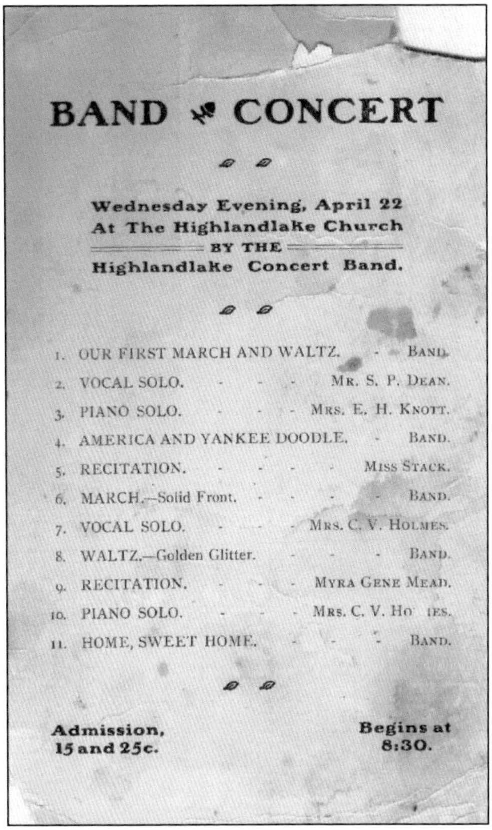

BAND CONCERT

Wednesday Evening, April 22
At The Highlandlake Church
BY THE
Highlandlake Concert Band.

1. OUR FIRST MARCH AND WALTZ. - - BAND.
2. VOCAL SOLO. - - - MR. S. P. DEAN.
3. PIANO SOLO. - - - MRS. E. H. KNOTT.
4. AMERICA AND YANKEE DOODLE. - BAND.
5. RECITATION. - - - - MISS STACK.
6. MARCH.—Solid Front. - - - - BAND.
7. VOCAL SOLO. - - - MRS. C. V. HOLMES.
8. WALTZ.—Golden Glitter. - - - BAND.
9. RECITATION. - - - MYRA GENE MEAD.
10. PIANO SOLO. - - - MRS. C. V. HO[illegible]ES.
11. HOME, SWEET HOME. - - - BAND.

Admission, 15 and 25c. **Begins at 8:30.**

This program for an evening of entertainment at the Highlandlake Church, dating around 1904 or 1905, is a fundraiser for the Highlandlake Concert Band, one of at least three bands in Highlandlake at the time. The back of the program has a handwritten recipe for an extremely high-sodium corn salad that would make most modern physicians gasp in horror. (Pauli Driver Smith collection, Historic Highlandlake.)

Rev. D.A. Strong served twice as pastor of the Highlandlake Church first from 1898 to 1901 and then from 1903 to 1907. He is shown here with nine women of various ages. Several are reading what appear to be hymnbooks, and one young woman is holding a guitar. None of the women are identified. (Patricia Jorgensen collection, Historic Highlandlake.)

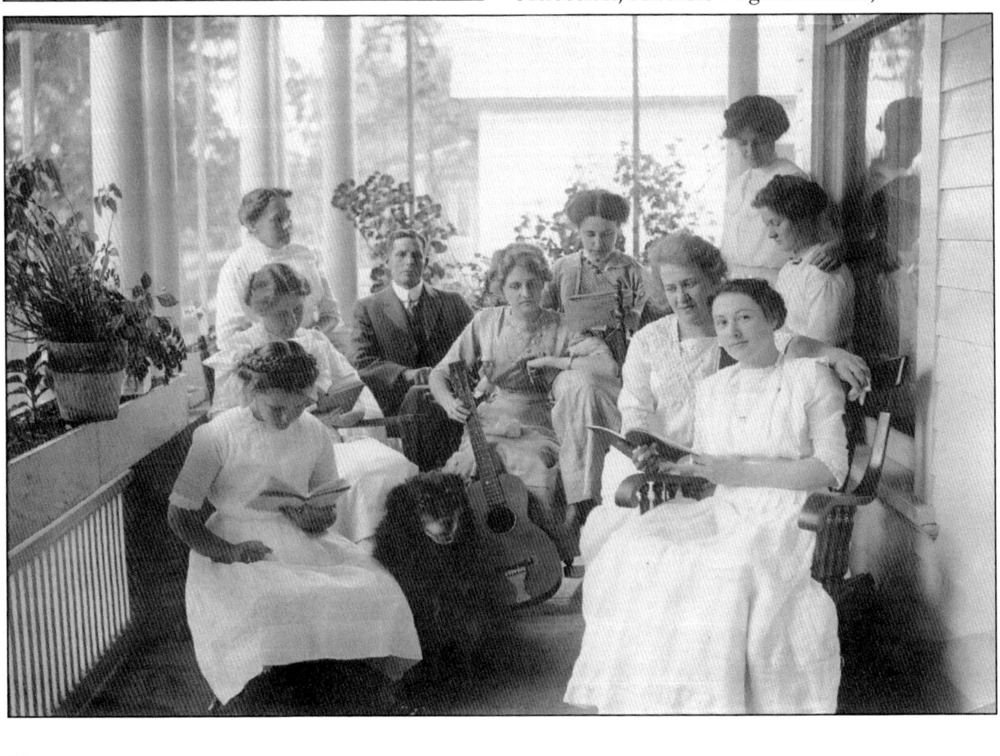

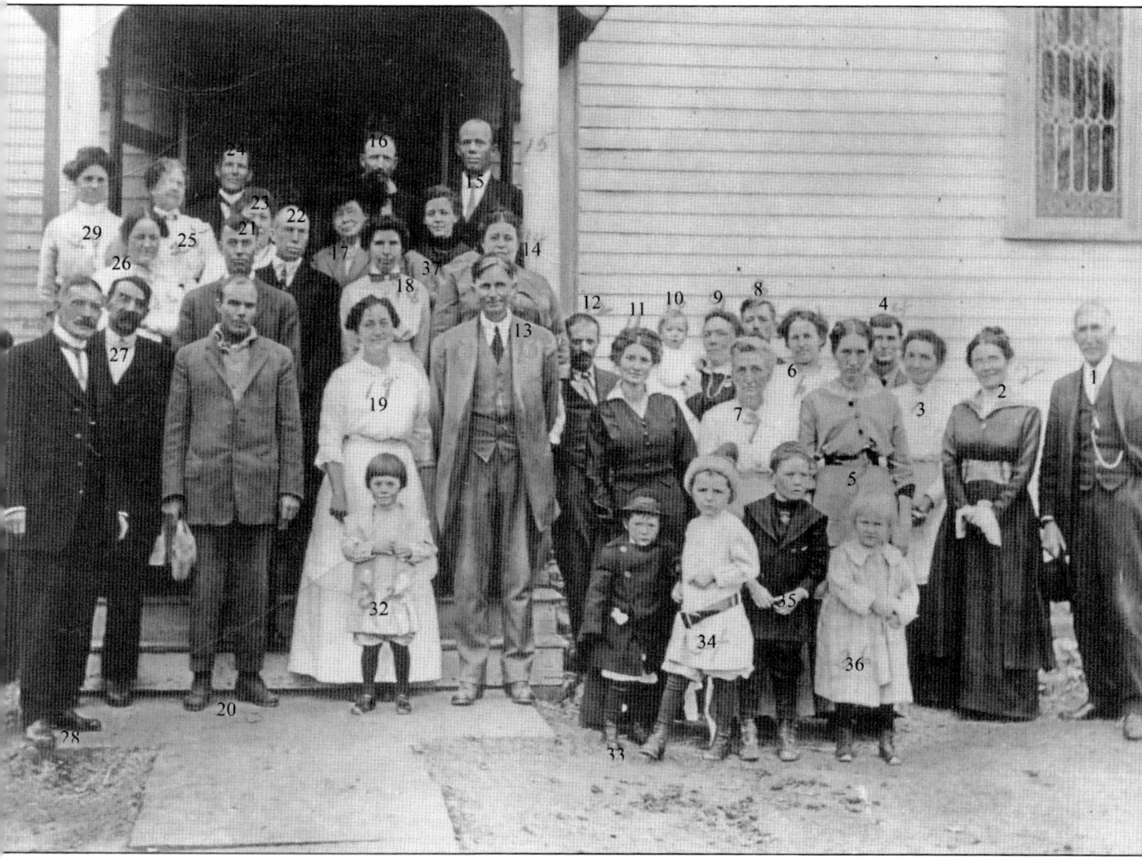

By 1914, there was only a handful of members attending the Congregational Church of Highlandlake. Those pictured are numbered as follows: 1. R.S. True, 2. Mrs. Emory, 3. Mrs. Jepperson, 4. Will Markham, 5. Mrs. Ollie Jones, 6. Ora Bishop, 7. Mrs. Hampton, 8. Fred Welty, 9. Mrs. Ed (Hallie) Kraig, 10. Helen Kraig, 11. Mrs. Walter (Caroline) Woodley, 12. Ed Kraig, 13. Andrew Backstrum, 14. Mrs. Backstrum, 15. Malcolm G. Mead, 16. F.H. Coates, 17. Mrs. D.A. (Ellen) Strong, 18. Mrs. F.H. (Grace) Coates, 19. Mrs. Clint (Menza) Smith, 20. Walter Woodley, 21. Carl True, 22. Ren True, 23. Mrs. Welty, 24. D.A. Strong, 25. Mrs. Bodkin, 26. Mrs. Will (Pearl) Markham, 27. Mr. Hampton, 28. Rev. C.P. Emory, 29. Mrs. Carl (Stella) True, 30. Mr. Woodley, 32. Kenneth True, 33. Frank Woodley, 34. Wayne Coates, 35. Frank Jepperson, 36. Marguerite Jones, and 37. Mrs. Malcolm (Florence) Mead. (Zola Coates collection, Historic Highlandlake.)

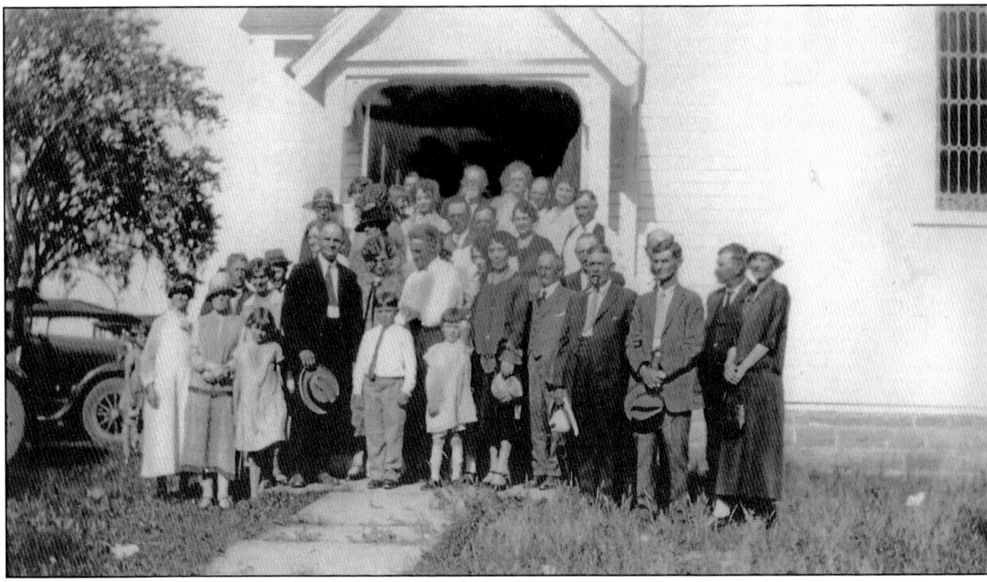

Over supper one evening in the spring of 1921, friends discussed having an annual pioneers' picnic and reunion in Highlandlake. They held the First Annual Pioneer Picnic in June 1921. This photograph is of the 1925 picnic, which over 150 attended. Ellen M. Strong, wife of the Rev. D.A. Strong, is standing on the far right. She died several months later. (Historic Highlandlake.)

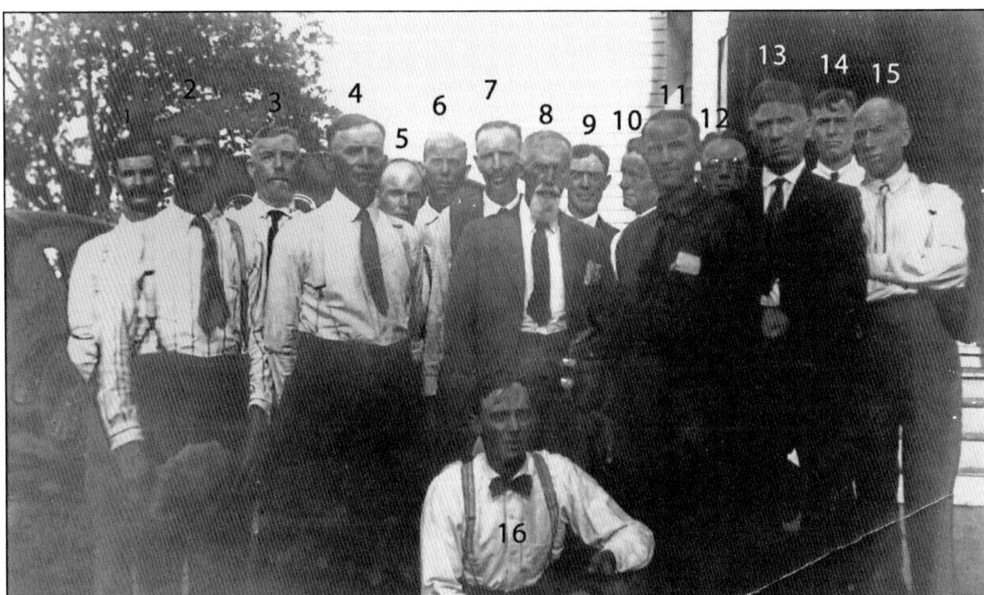

The 1929 Pioneer Picnic was also well attended. To commemorate the day, the photographer took pictures of several groups of people. In this group of men, only a few can be identified. They are numbered as follows: 1. Will Markham, 2. Walter Woodley, 6. Malcolm Mead, 7. Forbes Coates, 8. Charles Baker, 14. Carl True, and 15. Ren True. (Historic Highlandlake.)

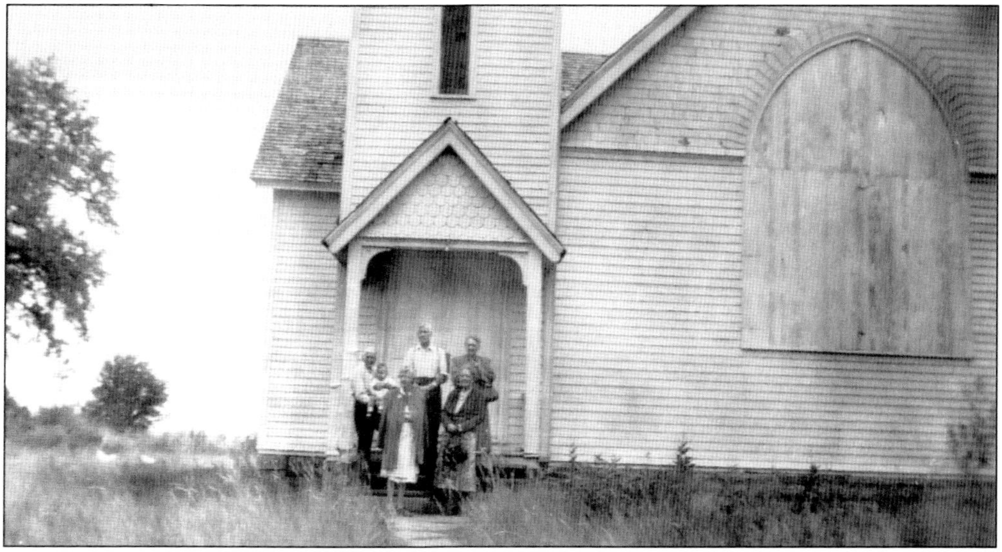

Former residents Emma Seybert and Gussie Holmes visited Highlandlake in the summer of 1949, and of course, a visit to the beloved old church was a "must do." The windows are shuttered to protect them from hail and vandals. Those pictured are, from left to right, (first row) Emma Holmes Seybert and Gussie Holmes; (second row) Carl True holding an unidentified infant, Ren True, and Florence Baker Mead. (Historic Highlandlake.)

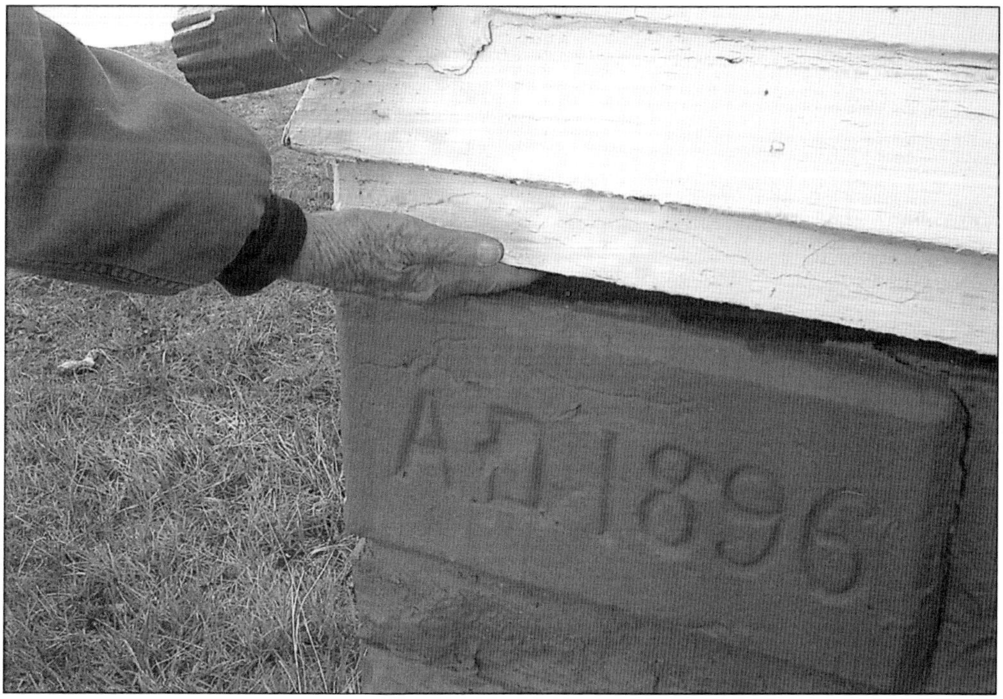

March 2003 brought the second-worst blizzard in Colorado's recorded history. Over three feet of snow, with drifts up to six feet deep, blanketed the Highlandlake Church building. The resulting weight shifted the foundation stones and walls. Saving the building took years of fundraising and grant writing. Here, Mary Mead Jensen slips her hand between the walls and the foundation, demonstrating just how serious the damage was. (Historic Highlandlake.)

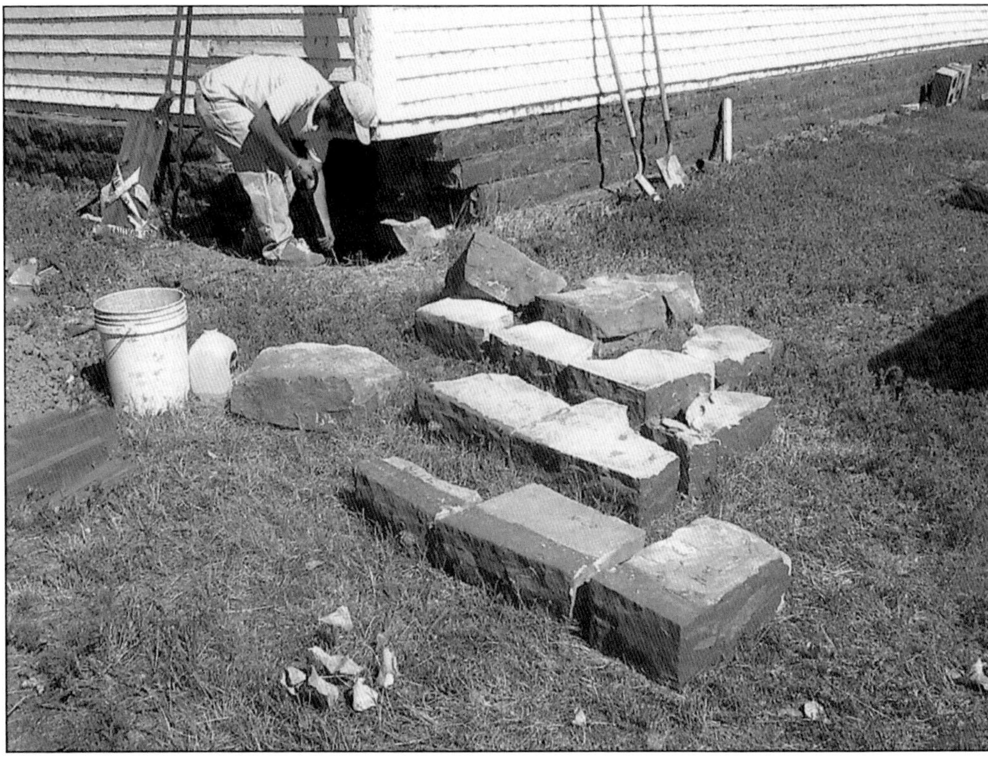

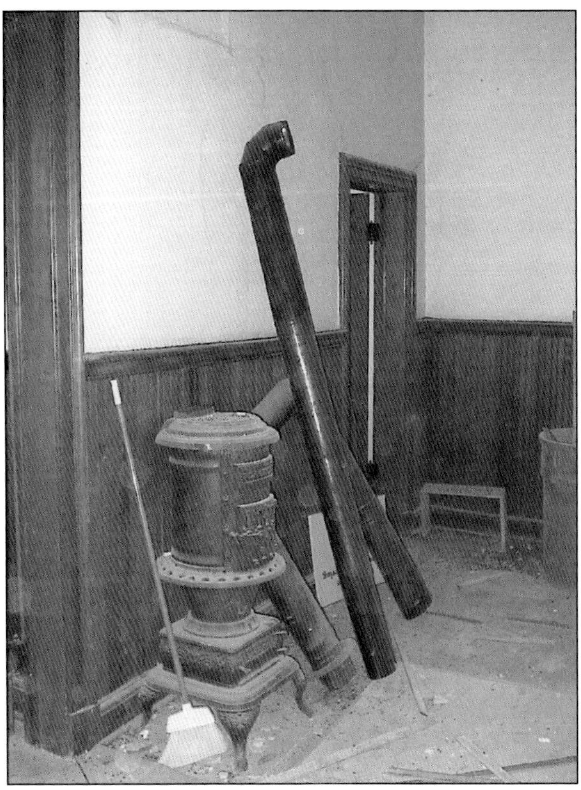

Restoration work, under the direction of architectural preservationist John Feinberg of the Collaborative, began in May 2003. In the first phase, workers carefully removed and numbered each foundation stone. After completing the new footings, they placed the stones back into their original positions. Based on chemical analyses of the original mortar done in Feinberg's lab, stonemason David Holliway perfectly recreated the original limestone mortar formula. (Historic Highlandlake.)

Damage went beyond the exterior of the building; the interior of the building was also heavily damaged. The moving foundation pulled the floors away from the walls, and large cracks appeared in both the walls and the stained-glass windowpanes. Here, the original potbelly stove is dismantled so it could be moved before the workers started plastering the walls. (Historic Highlandlake.)

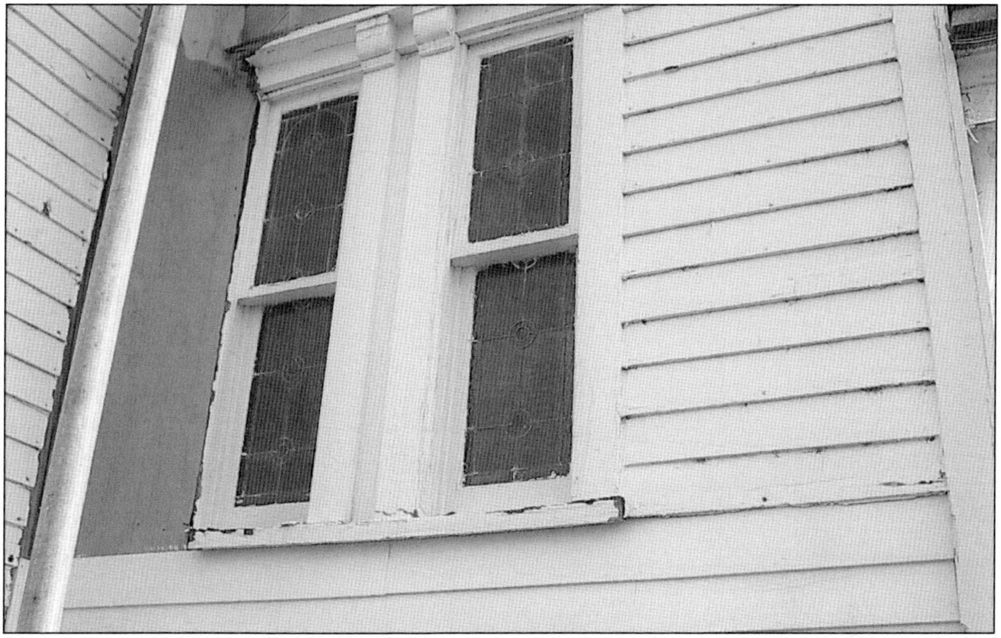

Extensive dry rot was discovered hidden behind the north and east faces of the bell tower. After widespread repairs, including reproducing the architectural window elements, reinforcing and straightening the tower, and adding a new drainage system, general contractor Peter Modafferi installed new siding. (Historic Highlandlake.)

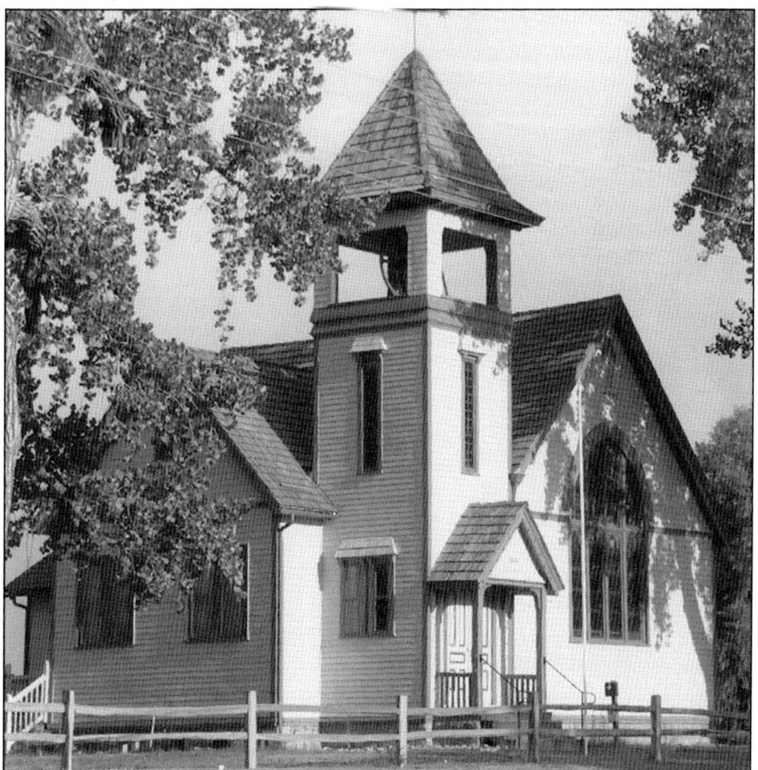

The project took over six years to complete. Hundreds of volunteer hours helped keep the costs under budget. The final task was to paint the building in its historic colors. Once again, John Feinberg's lab proved useful, this time in identifying the original color palette. Three main colors grace the exterior of this building, including butterscotch yellow with kelly and forest green trim. (Pauli Driver Smith.)

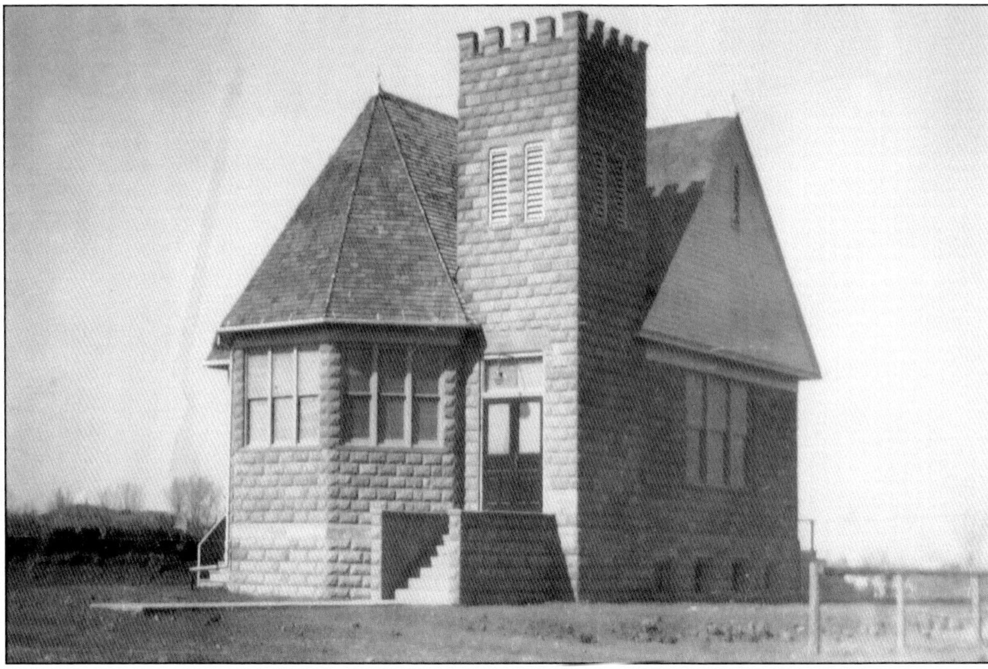

Mead United Brethren Church, now Mead Methodist Church, located on the corner of Fifth Street and Palmer Avenue, is the second-oldest church in the area. Organized in the early 1880s, the congregation met in the Mount Zion United Brethren Church near the intersection of Colorado Highway 66 and Weld County Road 7 until 1907, when the Mead church was built. (Neil and Linda Gehring collection, Historic Highlandlake.)

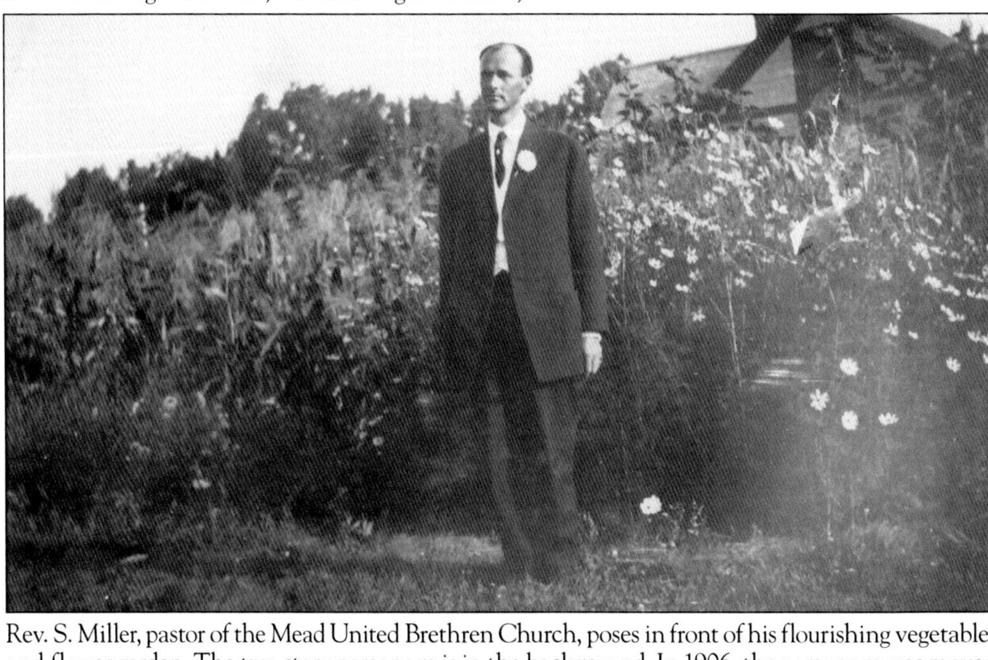

Rev. S. Miller, pastor of the Mead United Brethren Church, poses in front of his flourishing vegetable-and-flower garden. The two-story parsonage is in the background. In 1906, the parsonage was moved from the Mount Zion community to Mead. On February 13, 1919, a fire destroyed the building along with all the church records. (Mead Methodist Church collection, Historic Highlandlake.)

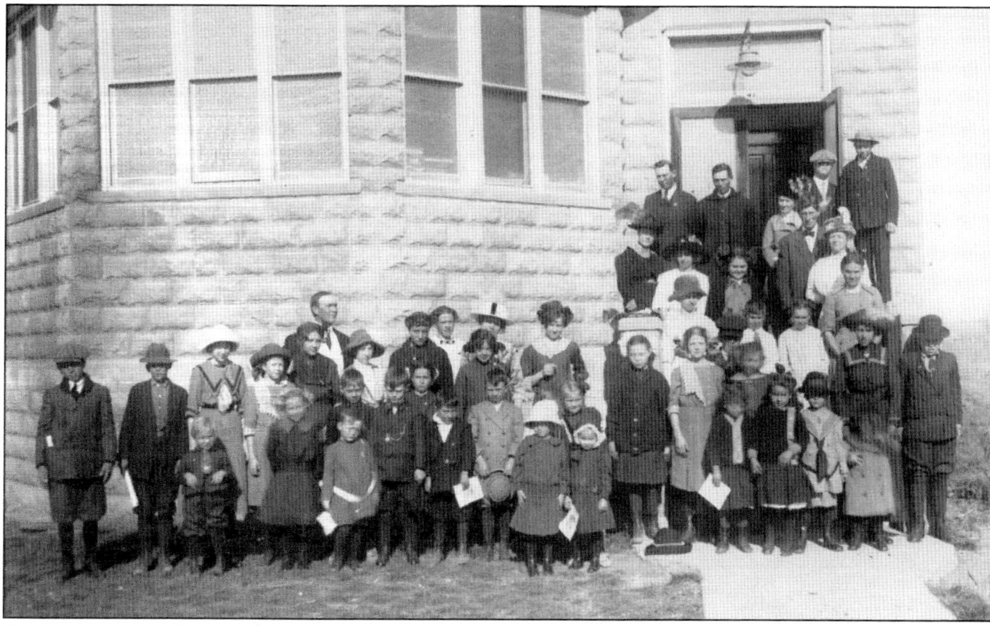

Sometime around 1917, the Mead United Brethren congregation gathered in front of its beautiful church. The occasion was probably the 10th anniversary of the move to Mead from Mount Zion. Trustee Benjamin Reese and his wife, Lydia, who taught Sunday school, are in the second row from the back on the right. Lydia is standing in front of her husband. (Mead Methodist Church collection, Historic Highlandlake.)

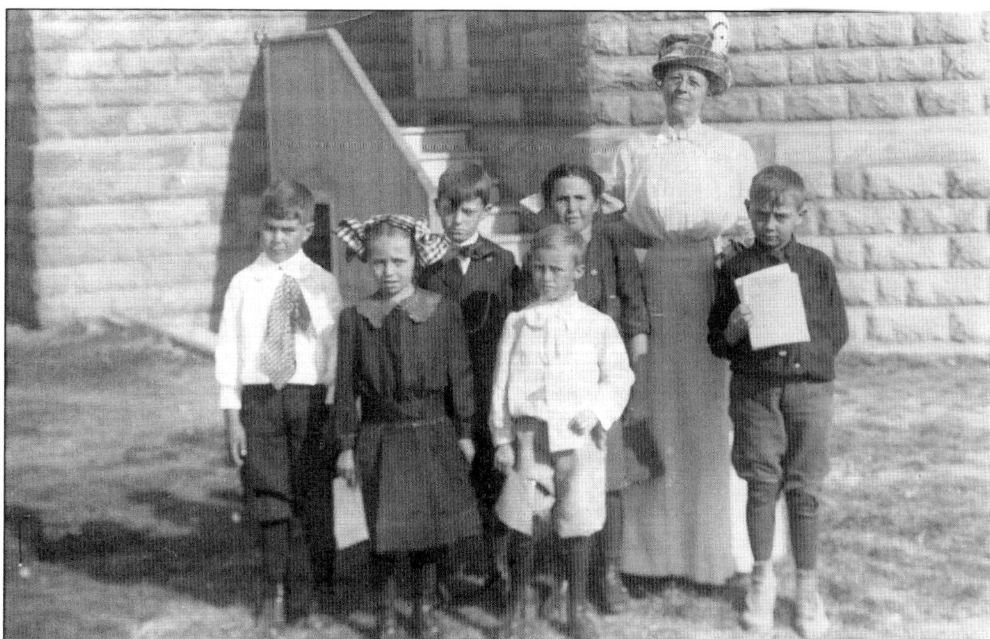

Lydia Reese poses near the side entrance to the Mead United Brethren Church with some of her Sunday school students. The building itself appears to be built of stone, but in reality, the blocks are made of molded cement. Several other homes in the immediate vicinity are also made of the same material. (Mead Methodist Church collection, Historic Highlandlake.)

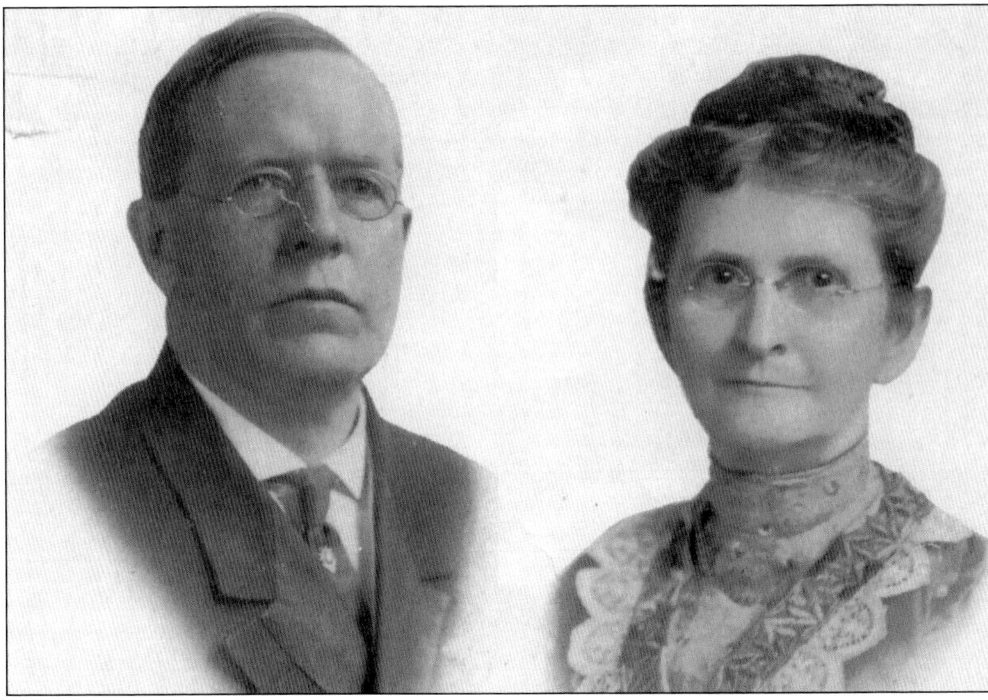

Benjamin Reese, who emigrated from Wales in 1885, was one of the first church trustees after the church moved to Mead. His wife, Lydia, taught Sunday school classes. They came to Colorado before 1900, settling in the Liberty-Mount Zion District. Both remained loyal members of the Mead church even after they moved to the Sunnyside District east of Berthoud and later to Longmont. (Mead Methodist Church collection, Historic Highlandlake.)

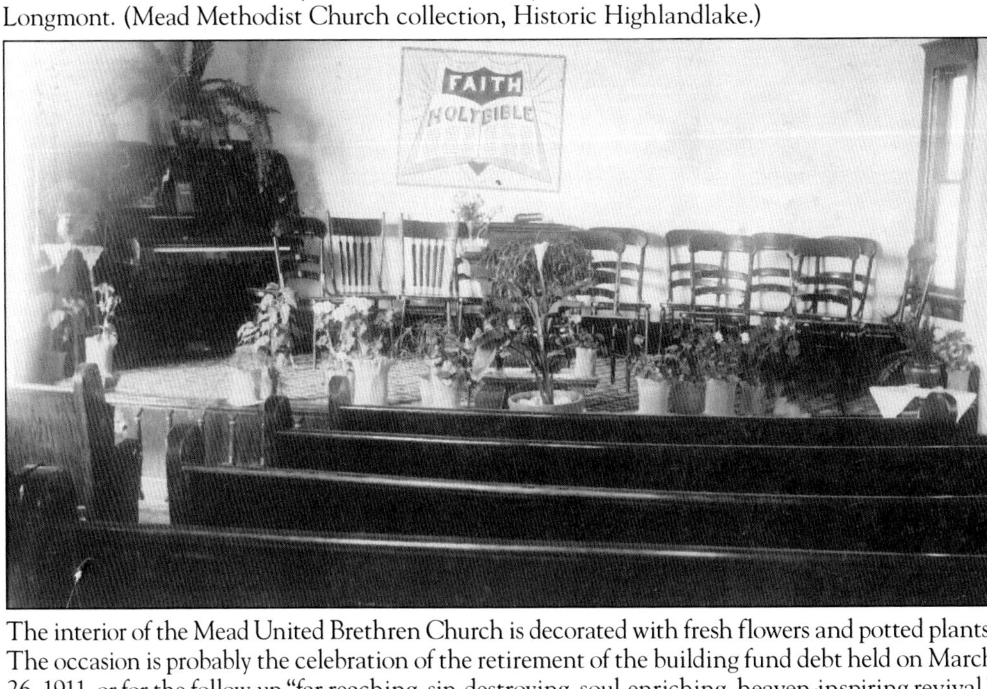

The interior of the Mead United Brethren Church is decorated with fresh flowers and potted plants. The occasion is probably the celebration of the retirement of the building fund debt held on March 26, 1911, or for the follow-up "far-reaching, sin-destroying, soul-enriching, heaven-inspiring revival," held later that same year. (Mead Methodist Church collection, Historic Highlandlake.)

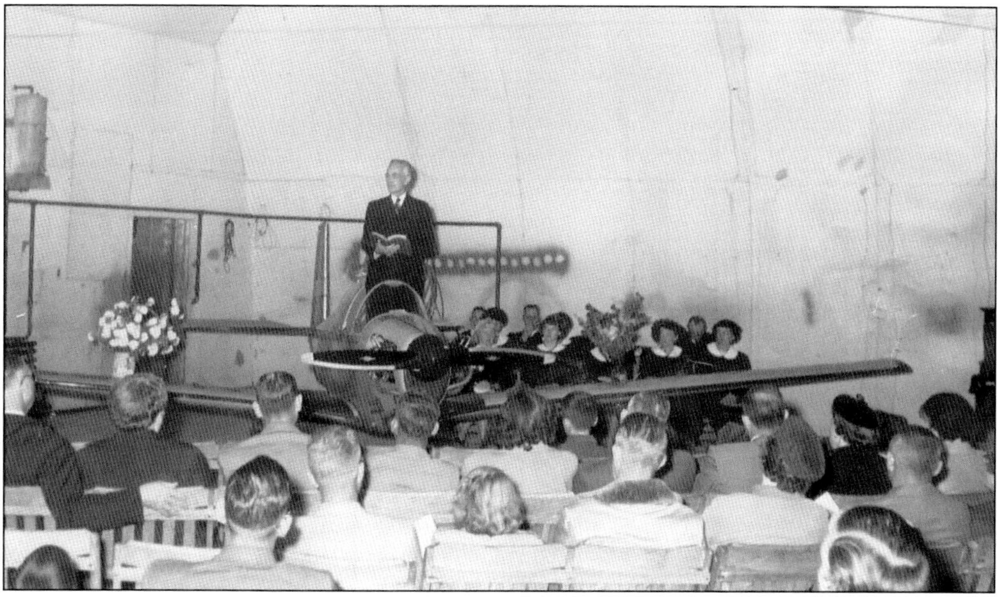

Rev. D.I. Hammond, pastor of the Mead United Brethren Church from 1942 to 1945 and from 1950 to 1952, had another passion, which was flying. As a member of the International Flying Farmers, he liked to refer to himself as "the flying preacher." In 1950, he preached a special sermon in the hanger at the Longmont Airport (now Vance Brand Airport) while standing in the cockpit of his plane. Below, Mrs. Wally (Gladys) Frederiksen accompanies the choir on the piano. Those pictured are, from left to right, (first row) Sally Peters, Alberta Humphrey, unidentified, unidentified, and Genove Brossman; (second row) Katherine (Hildreth) True and Barbara Fiechtner; (third row) Rodney Markham, Carl True, and Raimon Newby. (Both, Mead Methodist Church collection, Historic Highlandlake.)

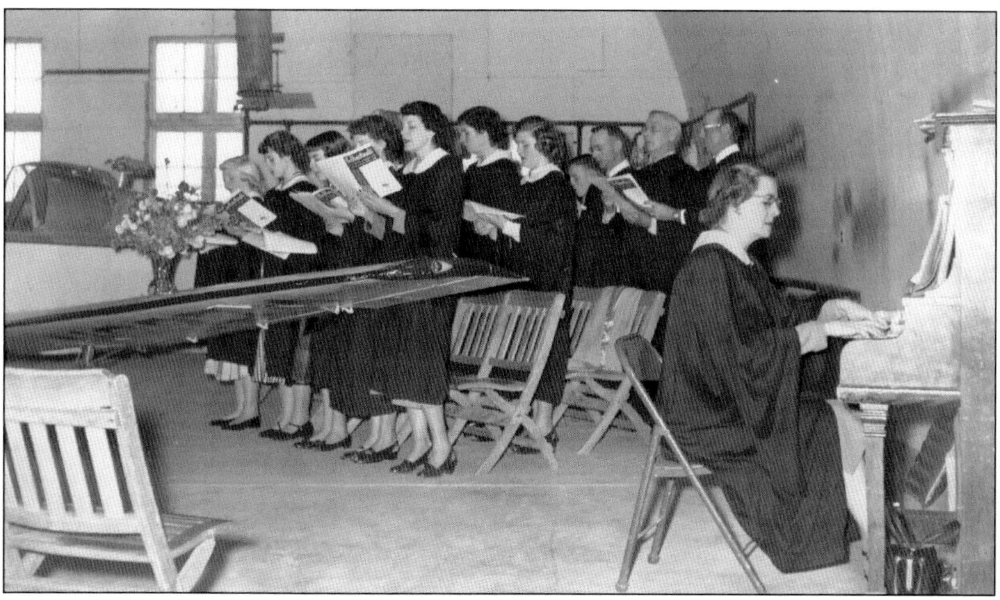

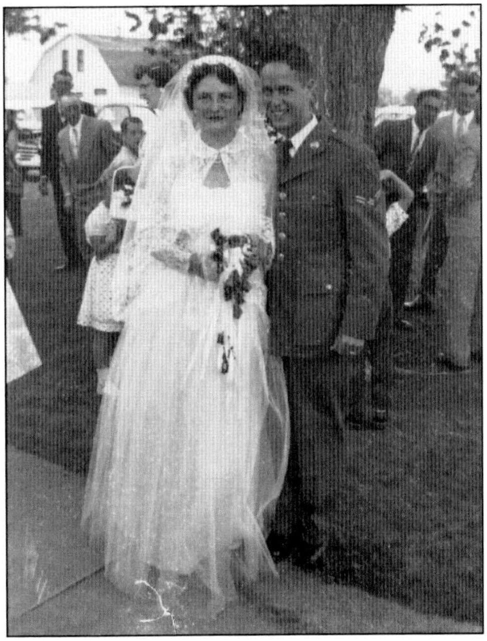

Weddings are always a popular celebration, and Walter and Barbara (Fiechtner) Logsdon's wedding in June 1955 was no exception. Over 120 guests attended the double-ring ceremony held at the Mead United Brethren Church. Walter was an airman second class stationed at Lowry Air Force Base, and Barbara, a graduate of Mead High school, was a nursing student at Denver's St. Luke's Hospital. (Walt and Barbara Logsdon.)

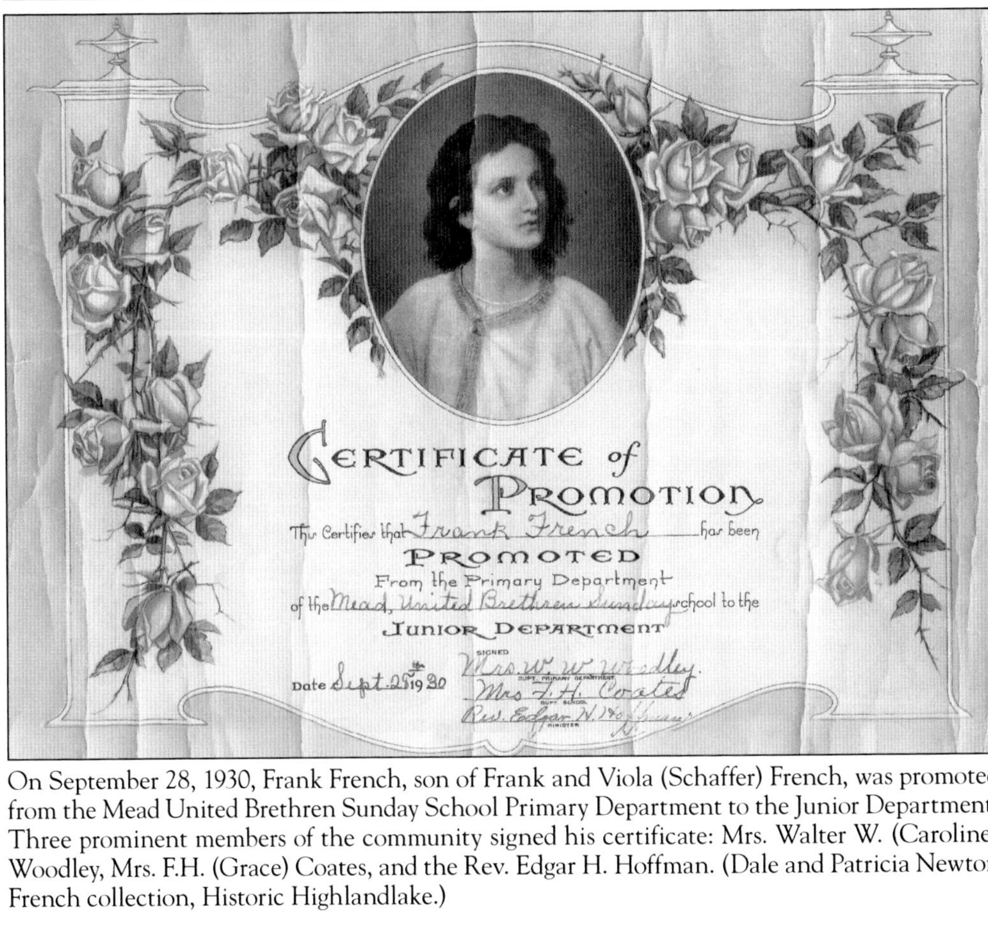

Certificate of Promotion

This Certifies that Frank French has been

PROMOTED

From the Primary Department

of the Mead, United Brethren Sunday school to the

Junior Department

Date Sept. 28th 1930

Signed

Mrs W. W. Woodley

Supt. Primary Department

Mrs F.H. Coates

Supt. School

Minister

On September 28, 1930, Frank French, son of Frank and Viola (Schaffer) French, was promoted from the Mead United Brethren Sunday School Primary Department to the Junior Department. Three prominent members of the community signed his certificate: Mrs. Walter W. (Caroline) Woodley, Mrs. F.H. (Grace) Coates, and the Rev. Edgar H. Hoffman. (Dale and Patricia Newton French collection, Historic Highlandlake.)

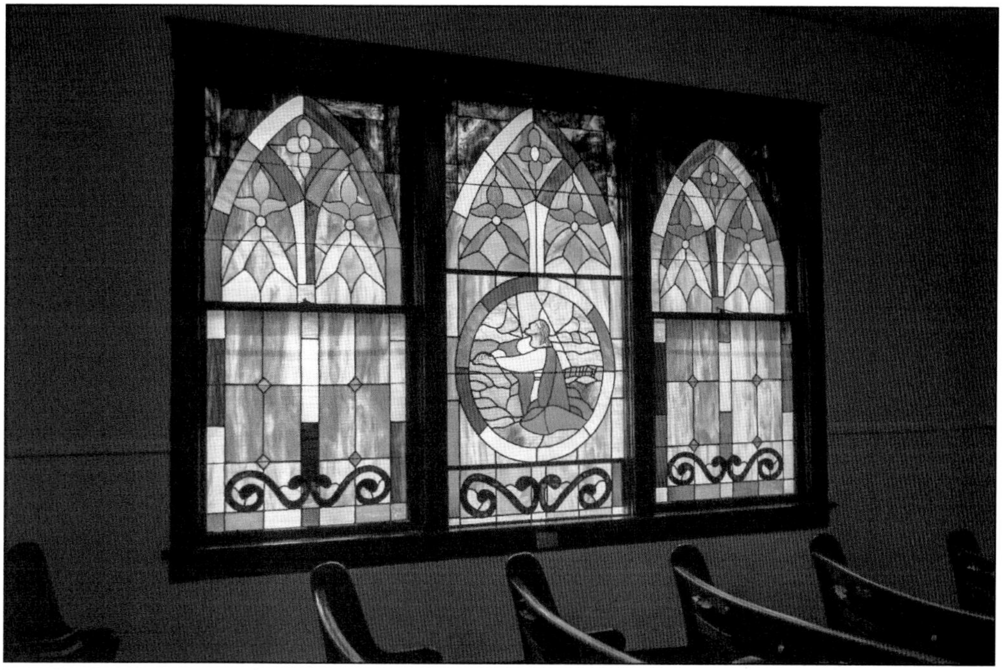

In 1979, Mary Peters, widow of Carl Peters, decided that the church needed stained-glass windows because, she explained, "I feel that they contribute to worship." As memorials to their husbands, she and Mrs. Edward (Lydia) Dahm, widow of the longest-serving pastor to that date, only planned to donate windows for the front of the church. That quickly changed; said Mrs. Peters, "We just got excited about our project and it snowballed until all the windows in the sanctuary were replaced." (Pauli Driver Smith collection, Historic Highlandlake.)

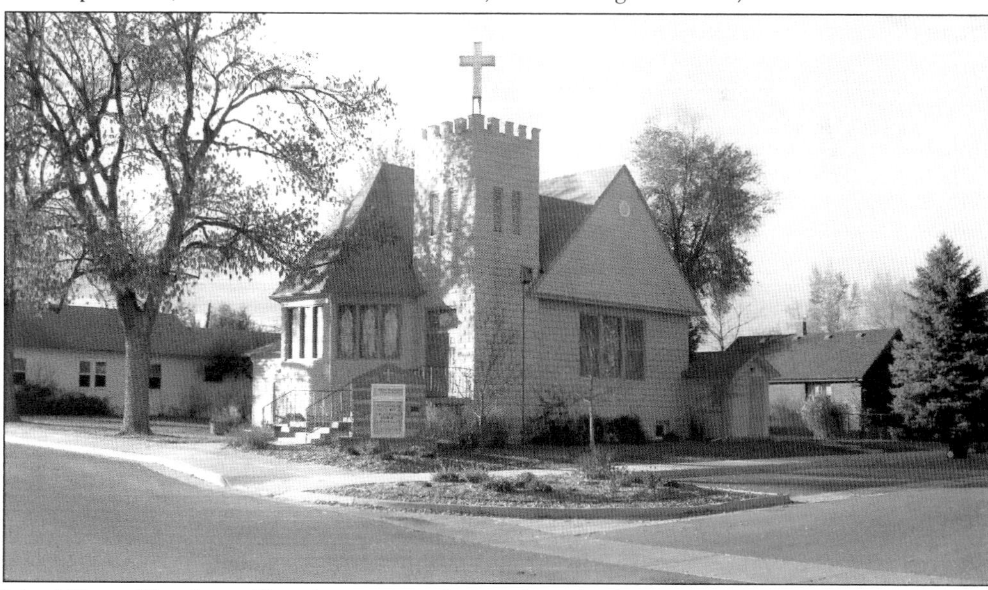

Mead United Brethren Church, now Mead United Methodist, is shown as it appears today. The prominent lit cross on the roof was donated by Carl A. Carlson in 1966 in memory of his wife, Ottilia (Olson) Carlson. To the left of the building is the parsonage built in 1919 to replace the one destroyed by fire. (Pauli Driver Smith collection, Historic Highlandlake.)

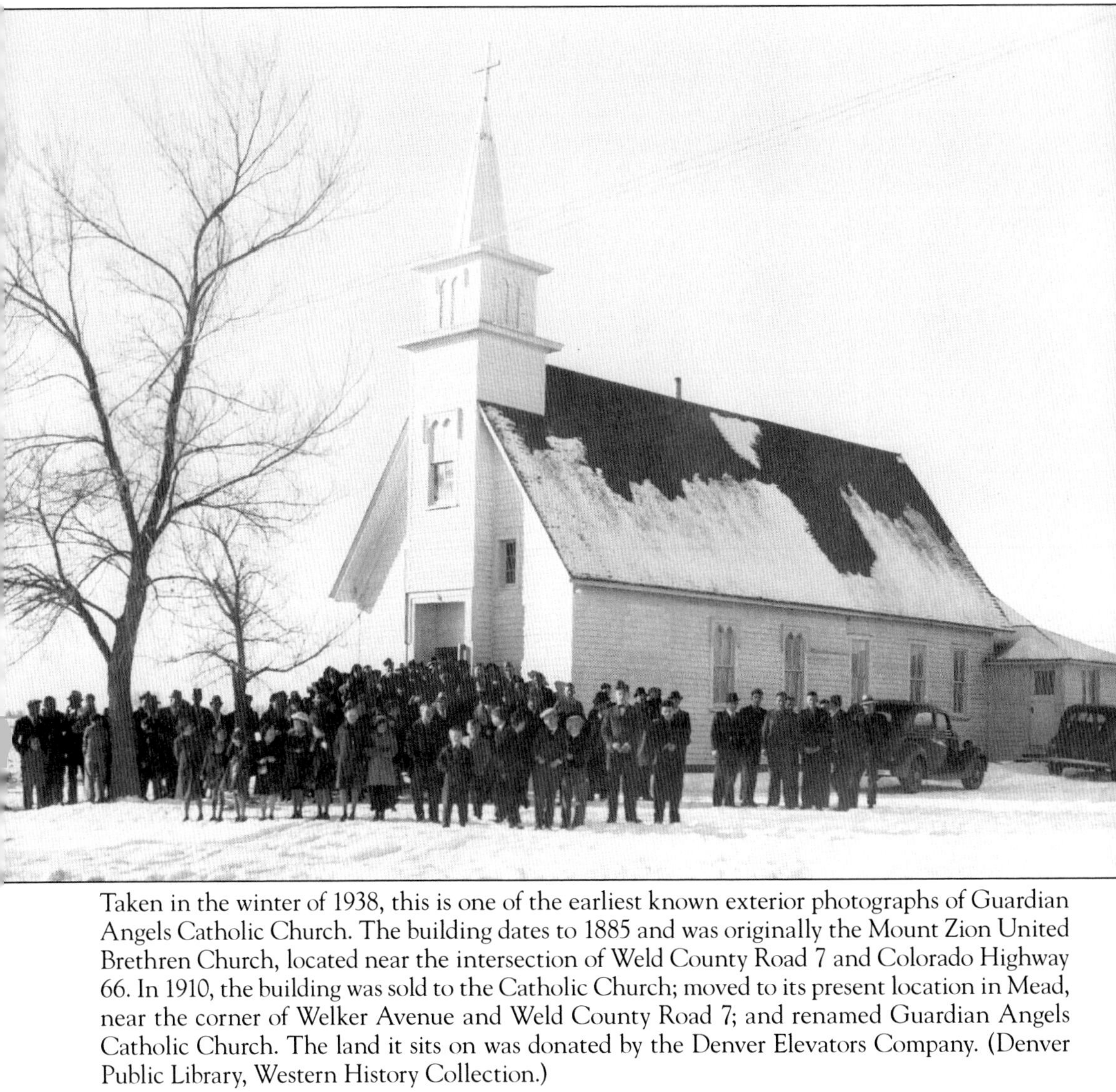

Taken in the winter of 1938, this is one of the earliest known exterior photographs of Guardian Angels Catholic Church. The building dates to 1885 and was originally the Mount Zion United Brethren Church, located near the intersection of Weld County Road 7 and Colorado Highway 66. In 1910, the building was sold to the Catholic Church; moved to its present location in Mead, near the corner of Welker Avenue and Weld County Road 7; and renamed Guardian Angels Catholic Church. The land it sits on was donated by the Denver Elevators Company. (Denver Public Library, Western History Collection.)

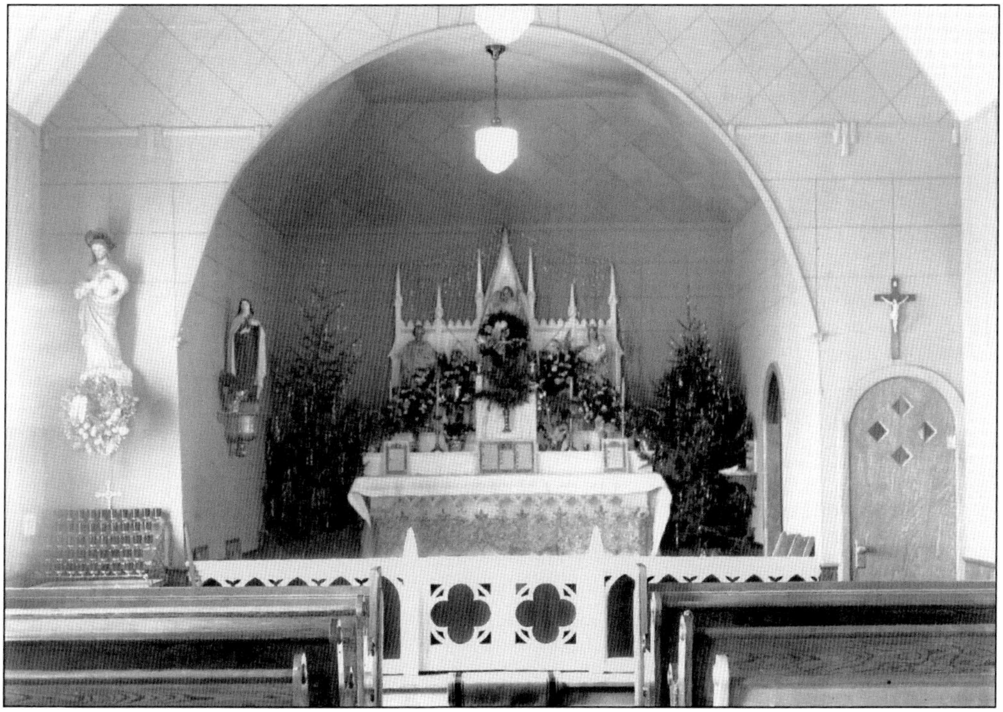

The first Mass held in the Guardian Angels Church was on January 1, 1911, with the Rev. Bernard Hamburger officiating. He continued to say Mass every other Sunday for five years until ill health forced him to retire. This photograph, taken around the date of that first service, commemorates that momentous occasion. (Archdiocese of Denver Archives and Guardian Angels Catholic Church.)

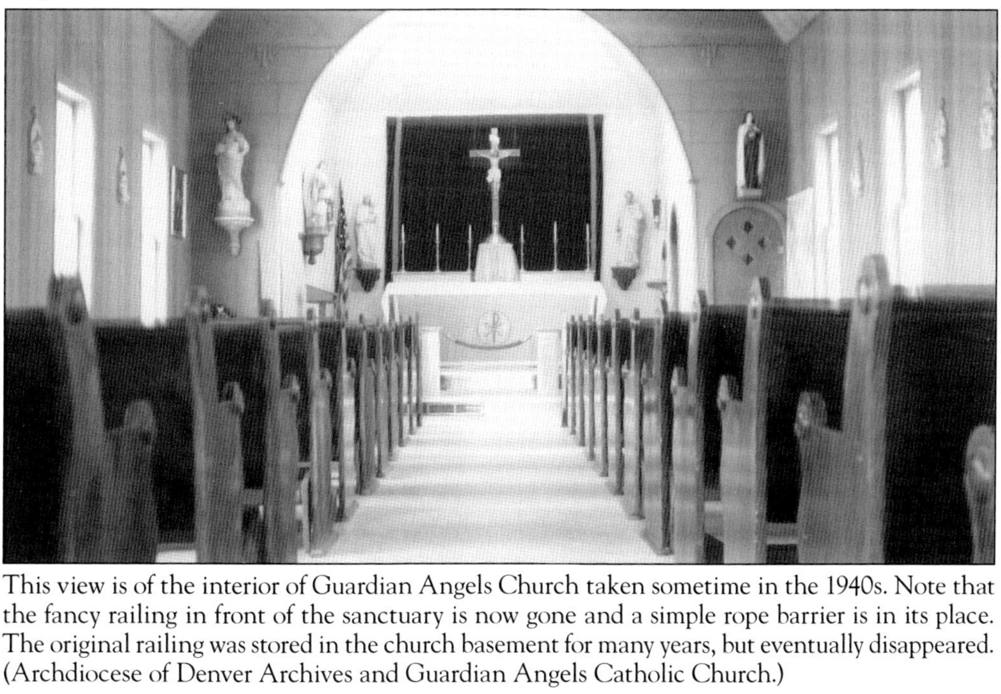

This view is of the interior of Guardian Angels Church taken sometime in the 1940s. Note that the fancy railing in front of the sanctuary is now gone and a simple rope barrier is in its place. The original railing was stored in the church basement for many years, but eventually disappeared. (Archdiocese of Denver Archives and Guardian Angels Catholic Church.)

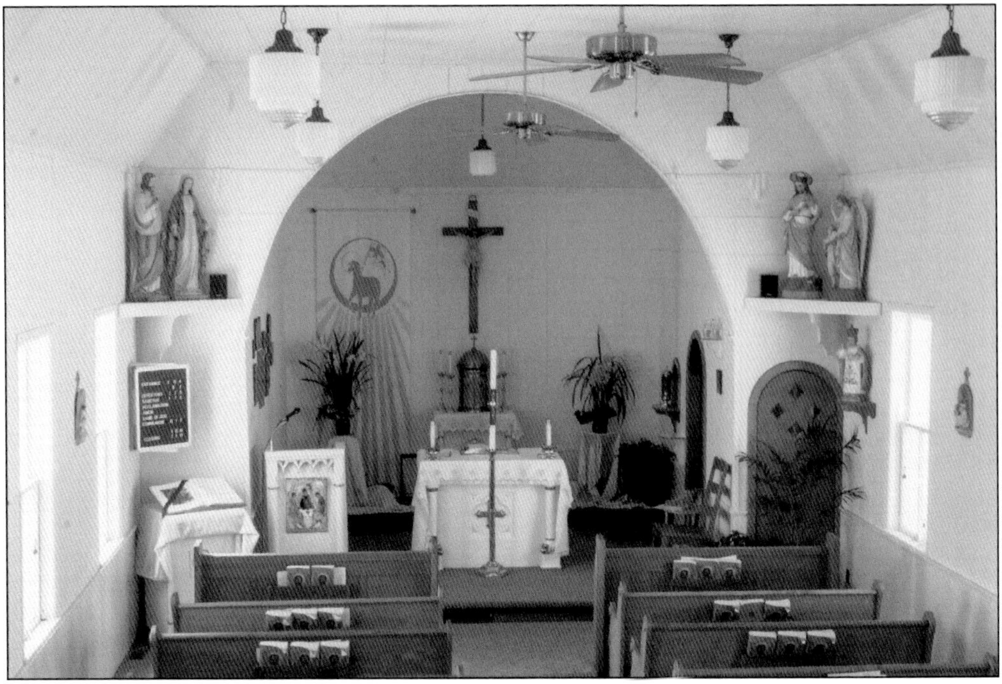

Here is a view of the interior as it looks today. There is now carpet on the floor, and the rope barrier to the sanctuary has been removed. Other changes include corner shelves to hold the statues, and ceiling fans and lighting are clearly visible. (Pauli Driver Smith collection, Historic Highlandlake.)

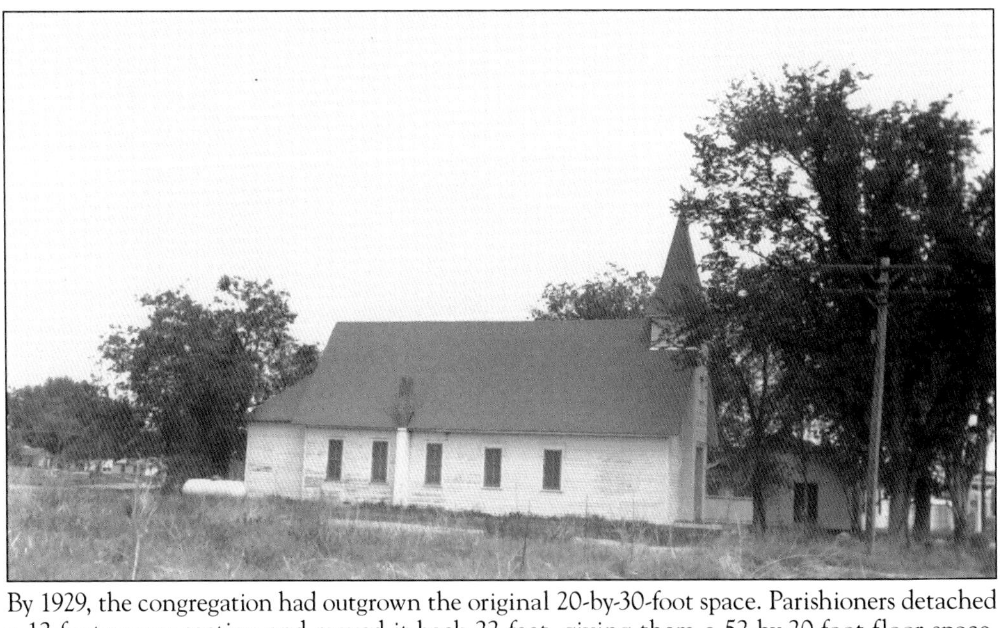

By 1929, the congregation had outgrown the original 20-by-30-foot space. Parishioners detached a 12-foot-square section and moved it back 22 feet, giving them a 52-by-20-foot floor space. The square section was reattached and used as the sacristy. If one looks closely at this c. 1962 photograph, the seams are where they inserted the new section. (Archdiocese of Denver Archives and Guardian Angels Catholic Church.)

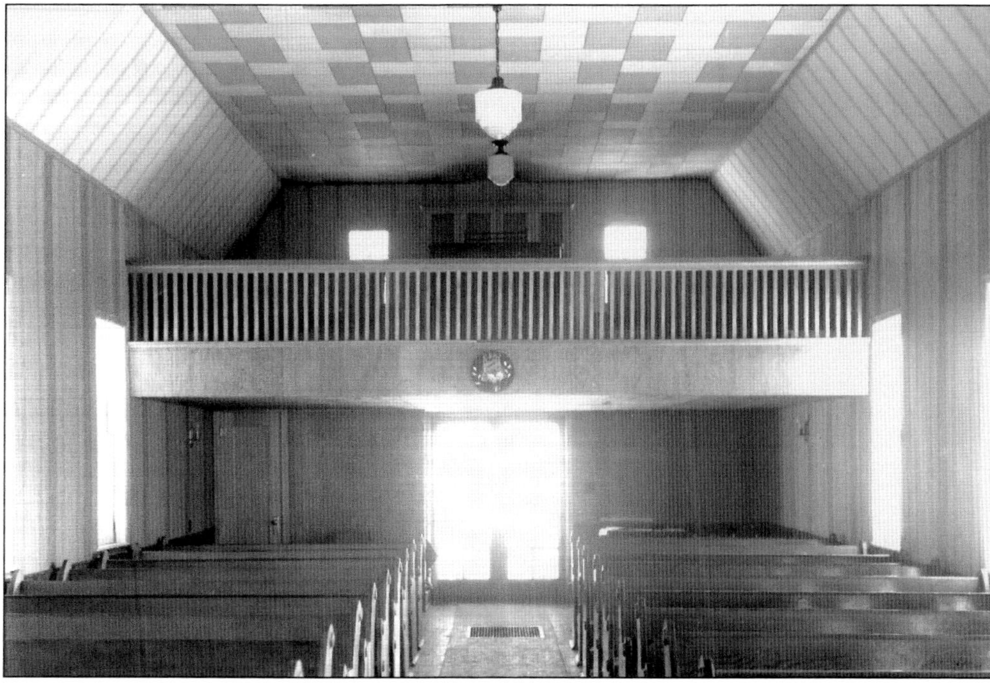

During the mid- to late 1930s, Fr. James Mahrer added a choir loft, accessed by a narrow staircase near the front entrance. To get the organ up into the loft, they had to hoist it up and over the front railing. (Archdiocese of Denver Archives and Guardian Angels Catholic Church.)

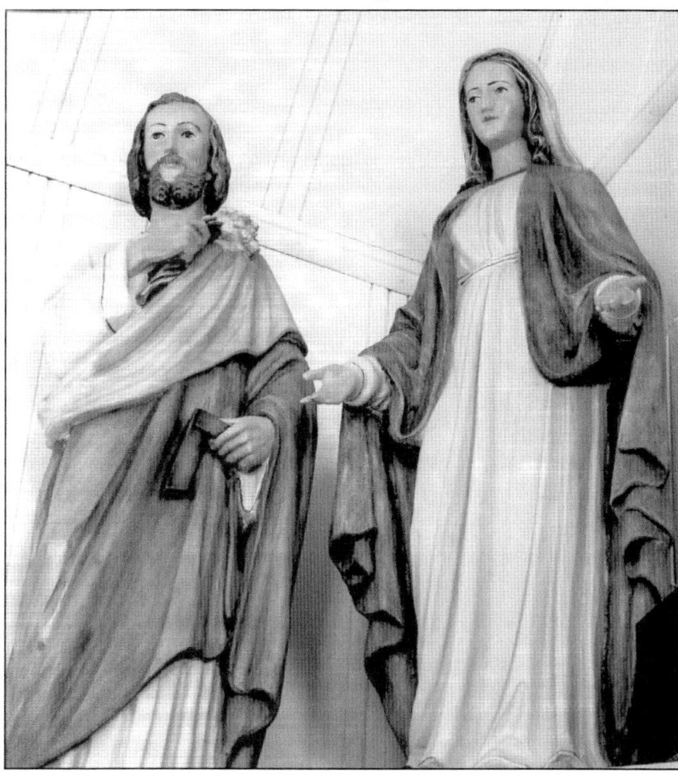

Statues of St. Joseph and the Virgin Mary are located on one of the corner shelves near the front of the room. The original grouping, while being restored in Denver, was lost, so the congregation replaced them with these. The original figures can be seen in the earlier interior photographs. (Pauli Driver Smith collection, Historic Highlandlake.)

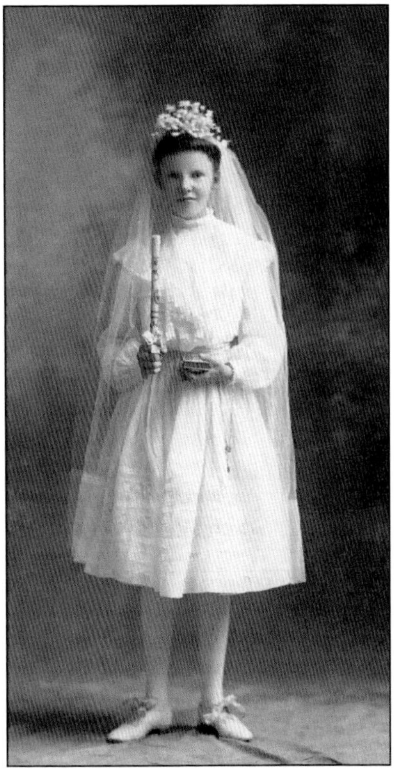

Traditionally, every church has certain rites of passage and milestones that members are expected to pass through. However, over the years, the recommended ages for these rites can change. For example, until the early in the 20th century, the Catholic practice was to administer the sacraments of Confirmation and First Communion after a child turned 12. However, by the 1920s, the recommended age for First Communion changed to seven or eight. Left, Ramona Mulligan, age 12, proudly poses for her First Communion portrait in 1892. Over 100 years later, seven-year-old Katie Rademacher, left, and an unidentified girl practice walking down the aisle for their First Communion. (Left, Mary Ann Mulligan collection, Historic Highlanlake; below, Patricia Tryon collection, Historic Highlandlake.)

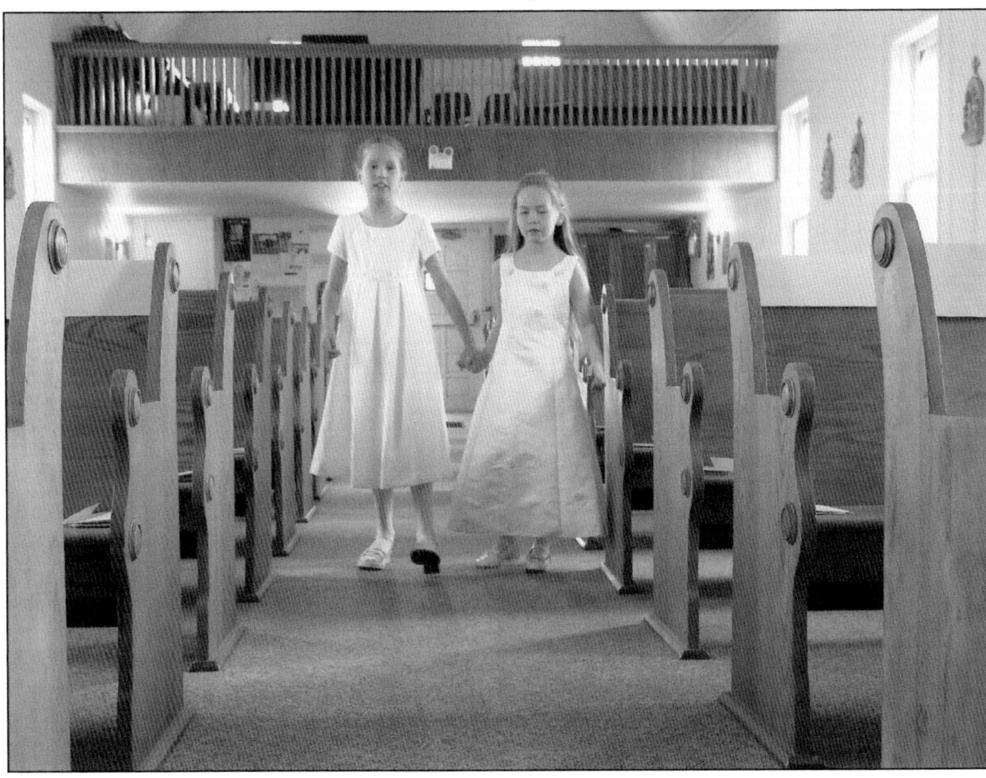

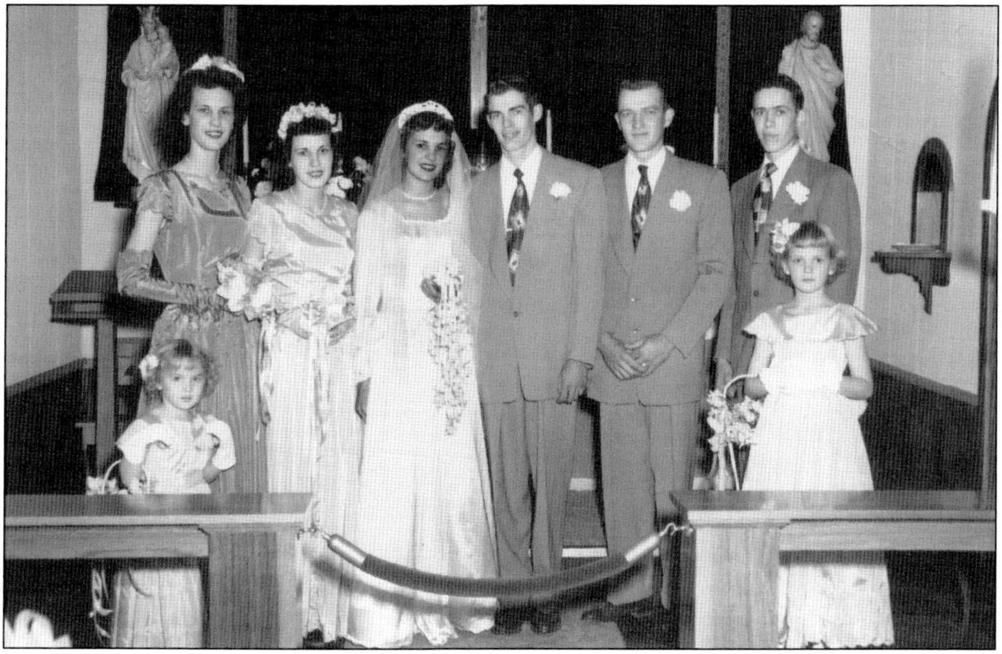

Weddings are another milestone usually marked by church ceremonies. Sunday, January 14, 1951, was the date that Patricia Ann Newton and Dale T. French chose to recite their marriage vows in the Guardian Angels Catholic Church. Shown above, from left to right, are maids of honor Rosemary Newton and Helen Newton; the bride and groom, Patricia Ann Newton and Dale T. French; best man Frank French Jr.; and groomsman Thomas Newton. Flower girls Sheri French and Diane Thornton are in front. Shown below are Larry French; Patricia Ann Newton; Helen Newton; Maureen Newton; and Rosemary Newton. (Both, Dale and Patricia French collection, Historic Highlandlake.)

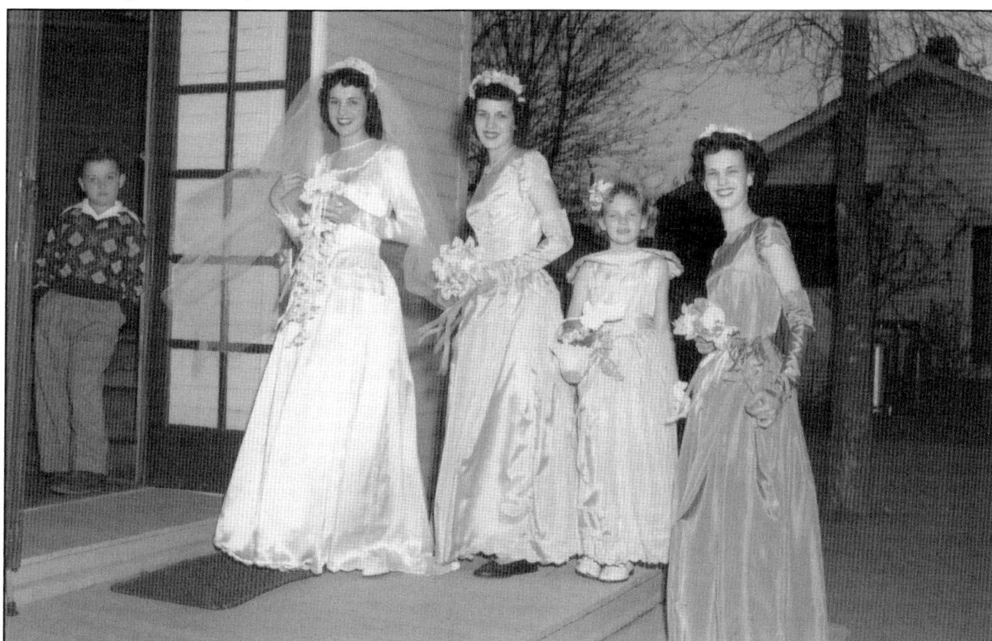

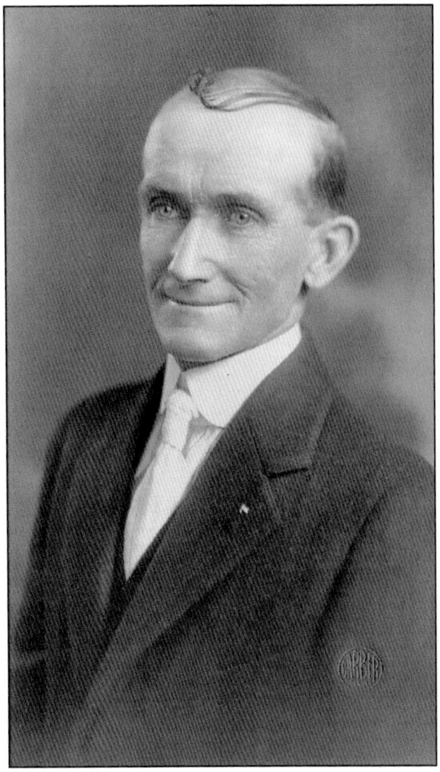

William and Martha Akers Redmond were prosperous Mead farmers and active members of the Guardian Angels Catholic Church. Their two daughters Alma, born in 1891, and Mary, born in 1892, eventually chose the religious life. In addition to their daughters, they also adopted a son, Thomas Michael O'Brian, and served as very generous godparents to several Mead children. (Both, Dale and Patricia French collection, Historic Highlandlake.)

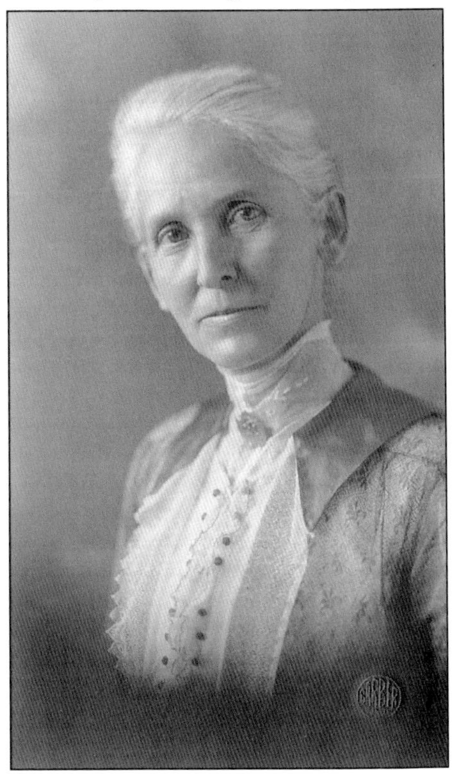

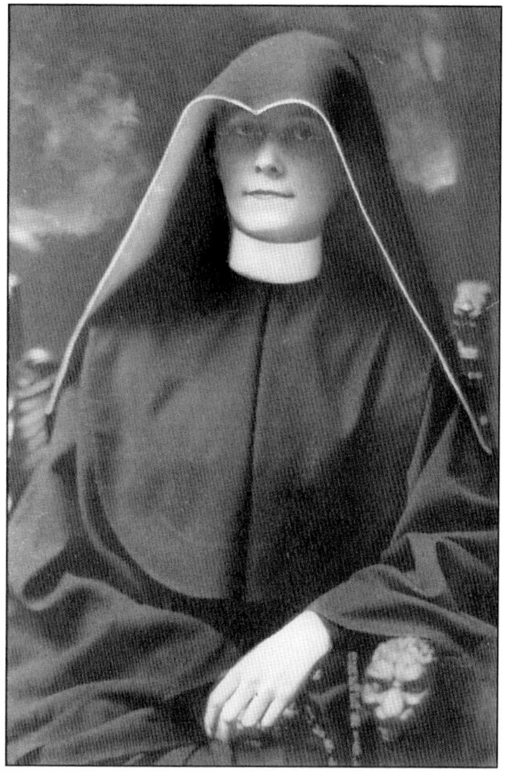

Shown are two beautiful hand-tinted calling cards for the daughters of William and Martha Akers Redmond. Sometime between 1910 and 1920, both girls joined the Sisters of Loretto Convent in Nerinx, Kentucky. Alma became Sister Mary Martha and Mary became Sister Pancratius. Both women eventually were assigned to teach at the Loretto Heights College in Denver. After retirement, they returned to the convent in Kentucky. Both are buried in the convent's cemetery. Rev. Charles Nerinckx founded the Sisters of Loretto religious community in 1812 with the goal of educating the poor children of the frontier. In the 200 years since the founding, the Sisters of Loretto have added chapters in 16 states (including Denver, Colorado) and 7 countries. (Both, Dale and Patricia French collection, Historic Highlandlake.)

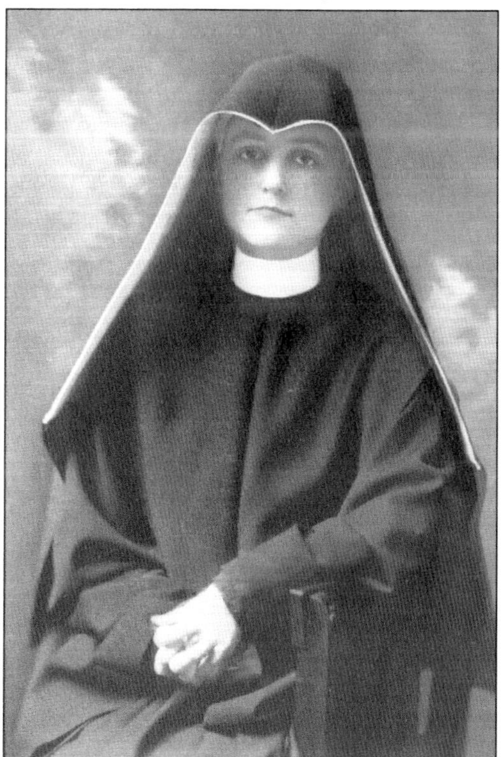

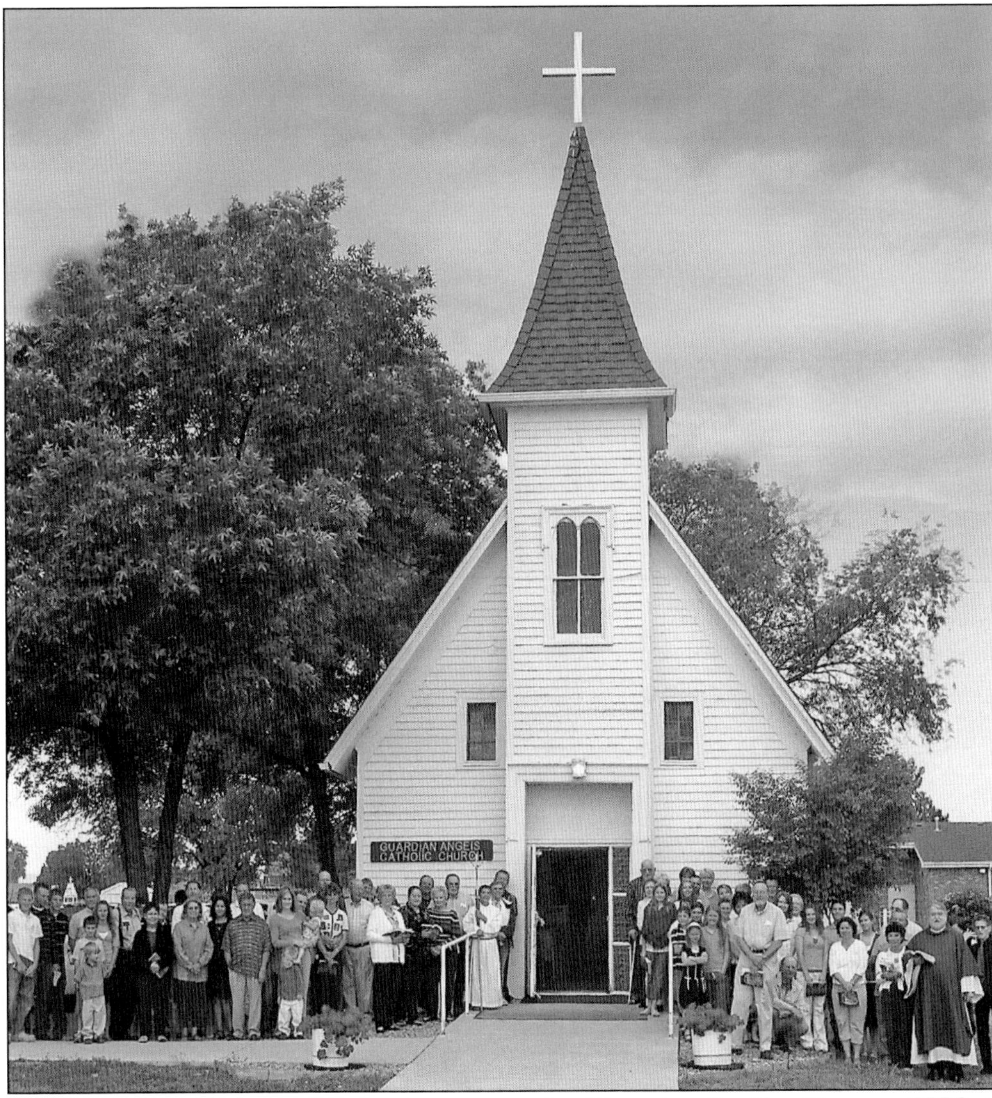

Guardian Angels Church is shown as it appears today. The priest, Fr. Alan Hartway, CPPS, is standing in the first row, far right. The altar boy standing with the cross to the left of the door is John Finke. Louis Rademacher (with cane) is standing back by the right side of the door. (Patricia Tryon collection, Historic Highlandlake.)

Six

Hollywood Comes to Highlandlake

Early one morning in November 1989, Pauli Smith's phone rang. She was still half asleep, so when the voice on the other end of the line asked if she would like to see Bruce Willis come to Highlandlake to film a movie, she thought it was a prank call. Rick Iverson from the Northern Colorado Film Commission assured her that it was not a joke and that the Highlandlake Church building was one of three area churches being considered for Willis's latest movie, *Die Hard II*.

Eventually, 20th Century Fox declared the church and lake "perfect," and the first week in January 1990, crews arrived to start work on the set. For three weeks, carpenters altered the appearance of the building by adding a false front and second bell tower to the back of the church, a picket fence around the perimeter, and a shed to house equipment including the snowmobiles used in the movie.

After weeks of prep work, the cast, filming crew, supporting people, and equipment arrived. Numerous newspaper articles documenting every move filled local and regional papers; the television show *Entertainment Tonight* interviewed the cast and even a few local residents; the movie lights lit up the sky for over 50 miles in every direction; and off-duty Weld County deputy sheriffs carefully guarded all entrances to Highlandlake.

Because no one except authorized personnel and actual Highlandlake residents were allowed into the closed set, it quickly became a game for people to try sneak in. Some managed to make it, but most were caught before they penetrated the outlying boundaries.

One of the perks of living on the lake was a brokered agreement that allowed residents to take as many personal pictures as they liked. The photographs in this chapter are gleaned from three such private collections. Since all the filming was done at night and no one had a professional camera, the quality is not always the best, but they give a small snapshot of a very cold but exciting time when Hollywood came to Highlandlake.

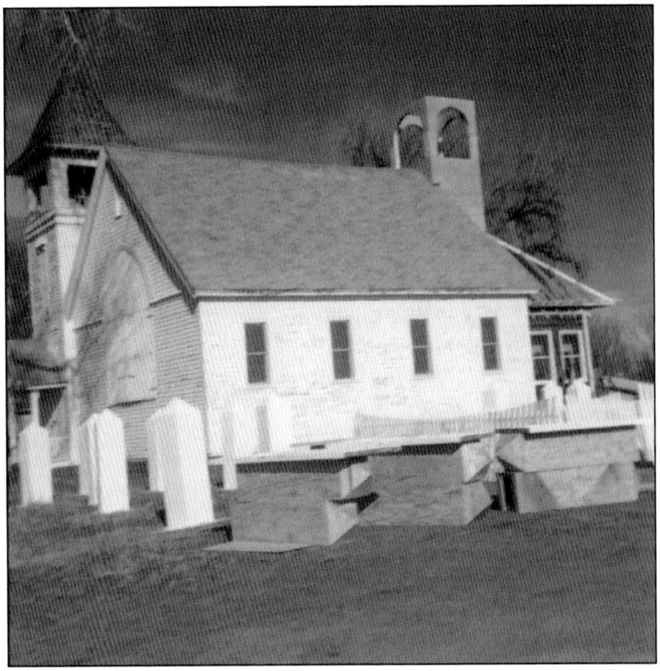

For three weeks before filming, crews built a false front on the back of the church and added a picket fence. On the right, one can see the new entrance and bell tower going up. The crew, being from Southern California, was not used to the cold. The supervisor sent them into Longmont to buy the warmest coats and boots they could find. (Mary Mead Jensen family collection, Historic Highlandlake.)

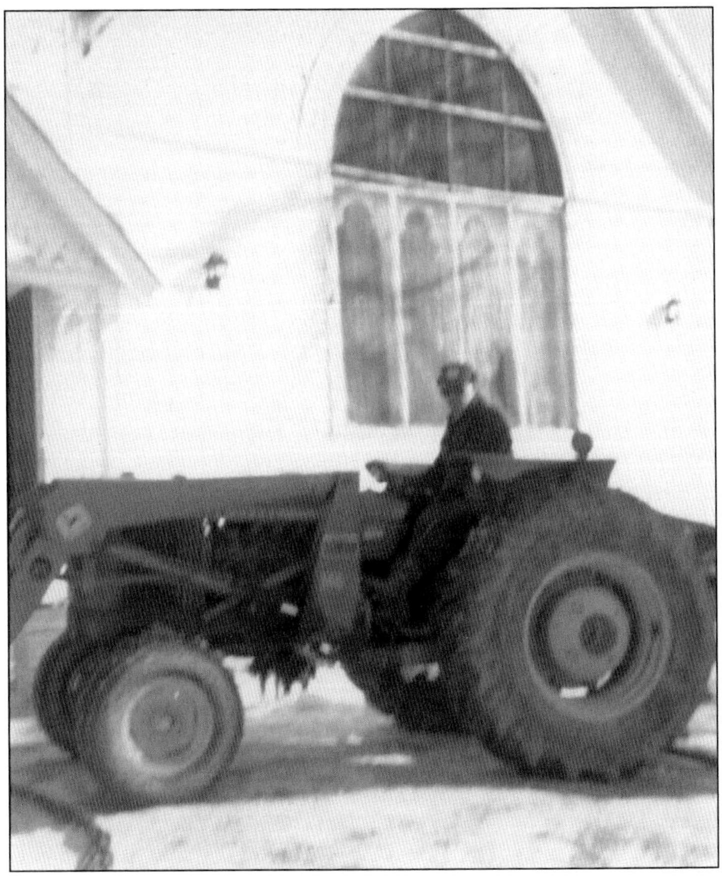

Locally rented equipment boosted the area's economy. It was estimated that 20th Century Fox spent somewhere between $100,000 and $150,000 in the Mead and Longmont area just for supplies, construction materials, rental equipment, items to decorate the sets, and other necessary expenses. (Mary Mead Jensen family collection, Historic Highlandlake.)

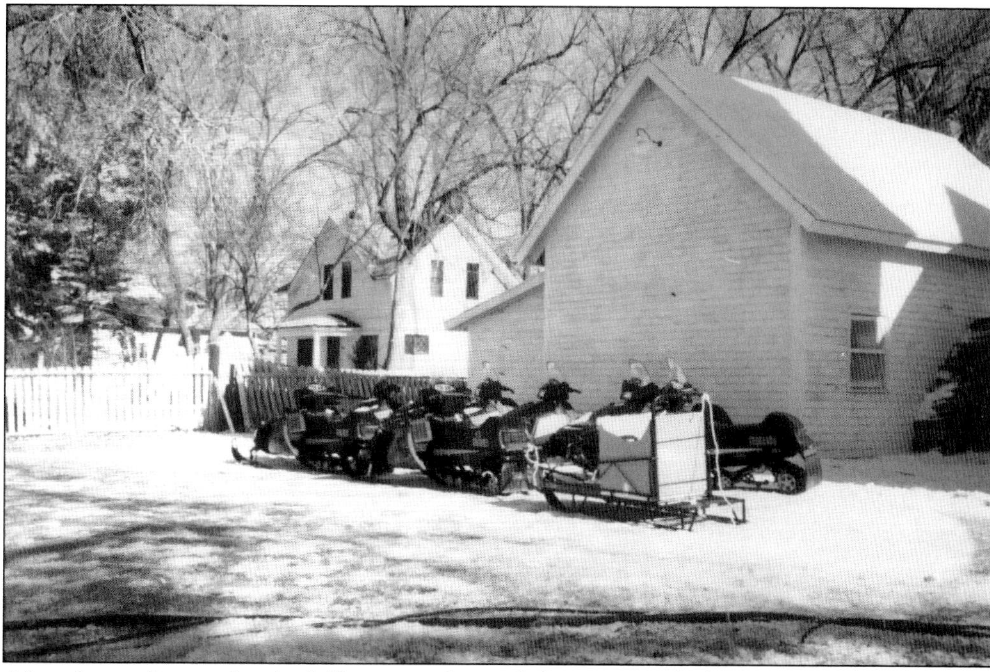

After the studio was finished filming, it donated this storage shed to Historic Highlandlake, Inc., which later auctioned it off as a fundraiser. Alongside the shed can be seen some of the snowmobiles, snow cats, and sleds, all of which were used during the filming. (Mary Mead Jensen family collection, Historic Highlandlake.)

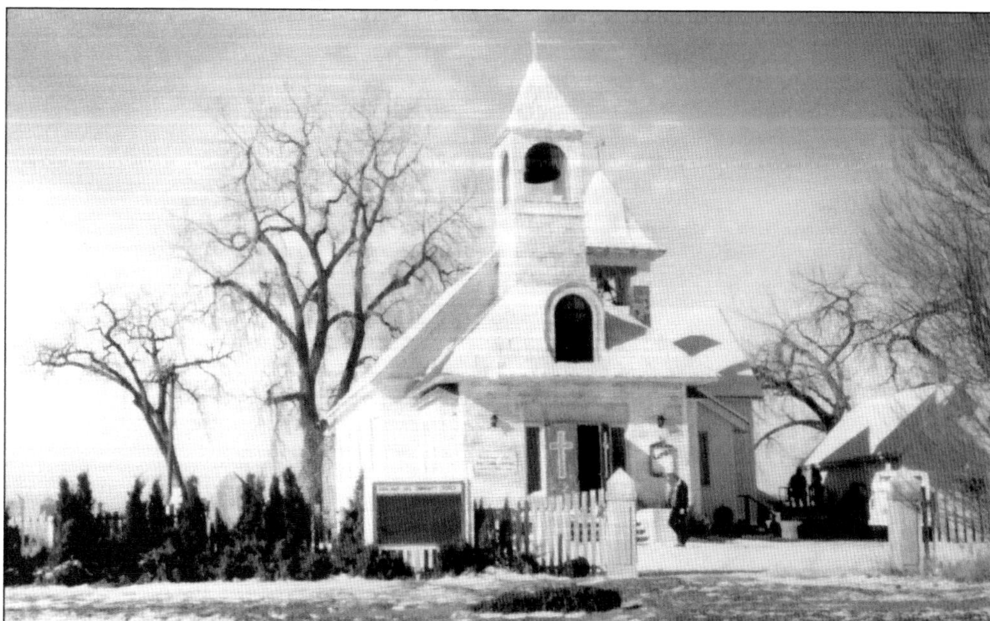

Here is a view of the new facade on the east side of the building. To the right can be seen the storage shed used to store equipment. On the left side, hidden by trees and shrubs rented from an area nursery, is the cemetery. Once filming started, the cemetery was removed to make room for the snow making machines. (Donna True King collection, Historic Highlandlake.)

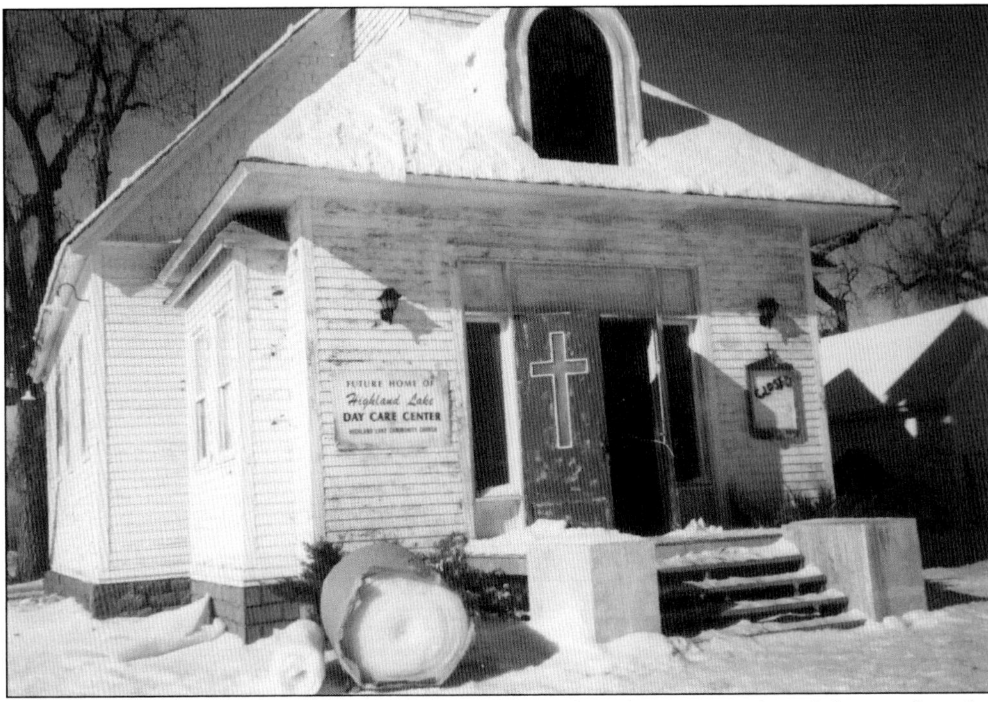

Above, workers stapled rolls of quilt batting to the roof as a base material, and then each night, huge blocks of ice were ground up, turned into snow, and sprayed on the roof and surrounding surfaces. The photograph below shows crews spraying snow on the roof, a process repeated each and every night just before filming. They also sprayed snow on the ground to cover areas disturbed by traffic and any melting from the night before. When finished, everything looked pristine, as if a snowstorm had just passed through. (Above, Donna True King collection, Historic Highlandlake; below, Brick House BBQ collection, Historic Highlandlake.)

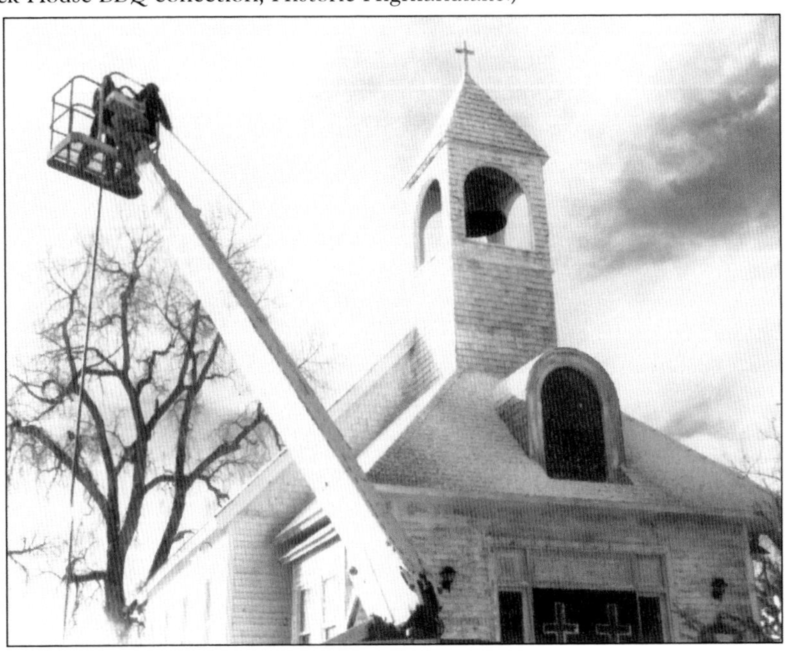

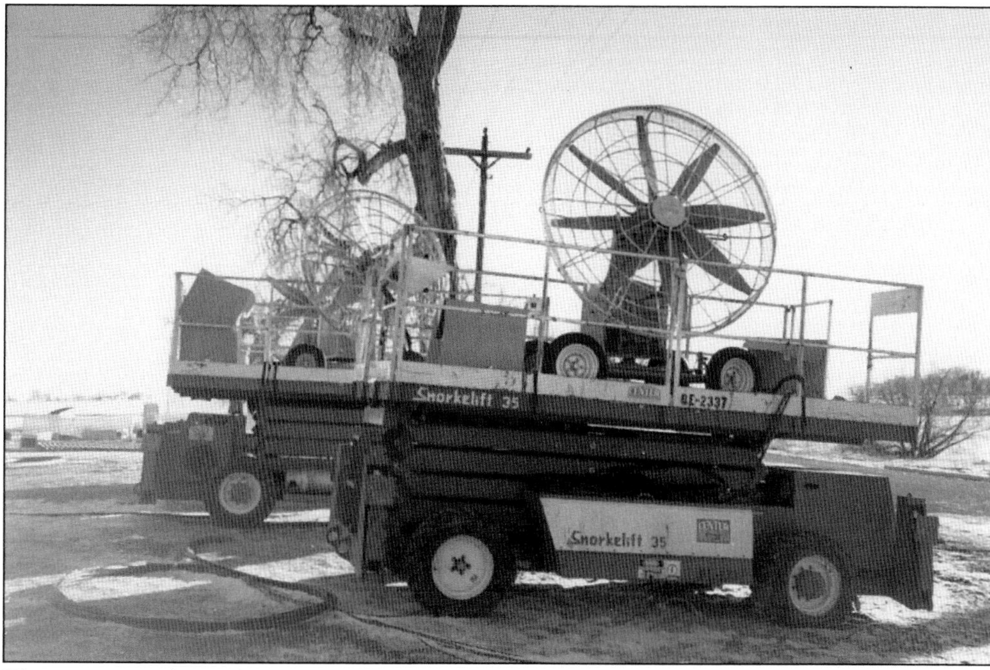

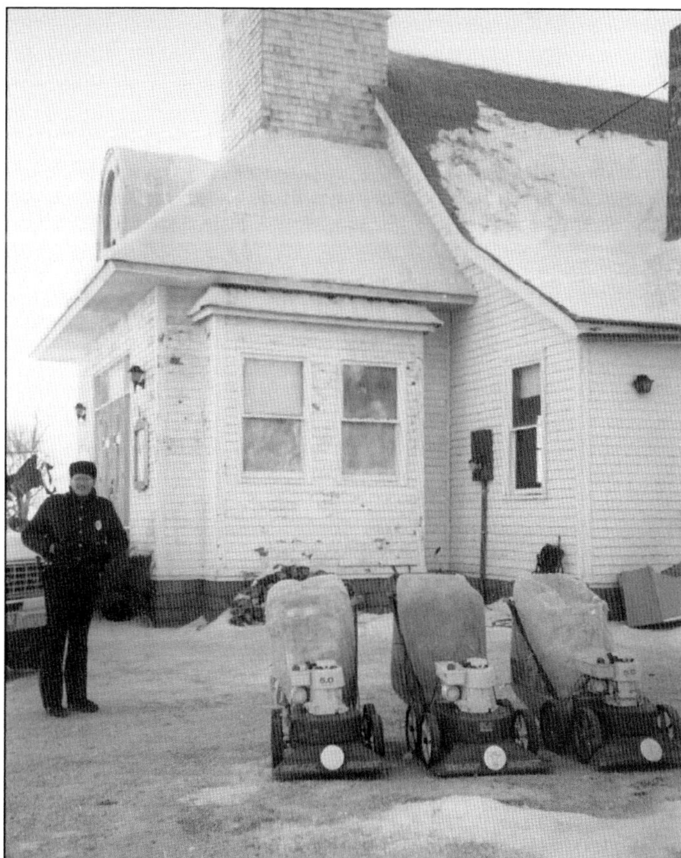

The main story line takes place at midnight on Christmas Eve during a blizzard. To simulate blowing snow, these large fans blew biodegradable, fake snowflakes toward the church. After filming was over, the studio hired locals to rake and vacuum up the fake snow. However, for almost a decade remnants of these snowflakes were still seen, clustered along the edges of the irrigation ditches and caught in the branches of low-lying bushes. At right are industrial-sized machines, which somewhat resembled lawnmowers, used to collect the fake snow into bags. One of the extras, who played a police officer, stands near the machines. (Both, Mary Mead Jensen family collection, Historic Highlandlake.)

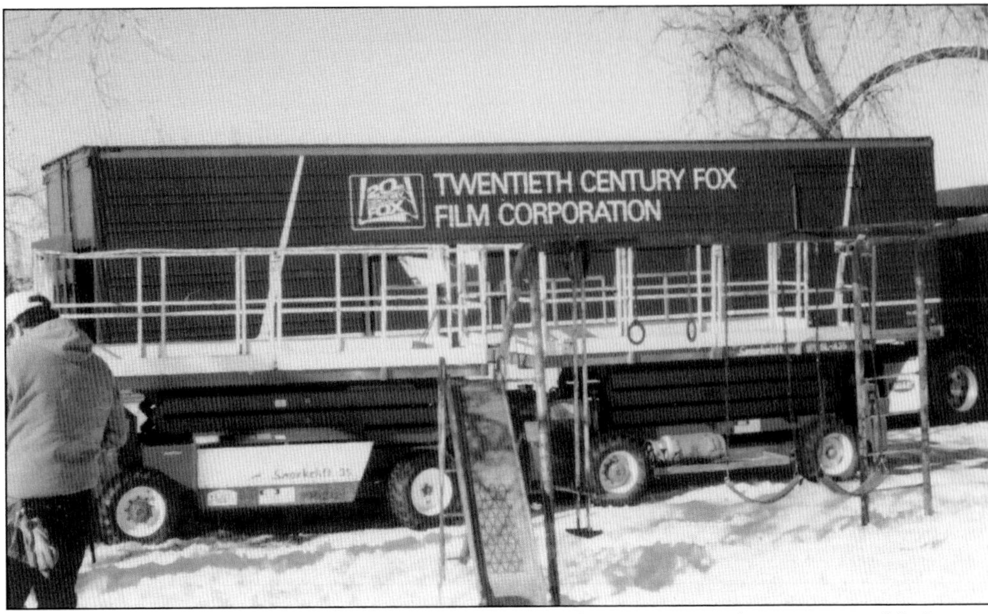

During the little over a week of filming, 62 vehicles—from snow machines to portable booms, semitrailer trucks, wind machines, portable dressing rooms, to power generators on wheels—lined the two roads through Highlandlake. Each evening, five busloads of actors, technicians, sound and light people, extras, directors, producers, key grips, and other movie folks arrived and set up shop for the night. (Donna True King collection, Historic Highlandlake.)

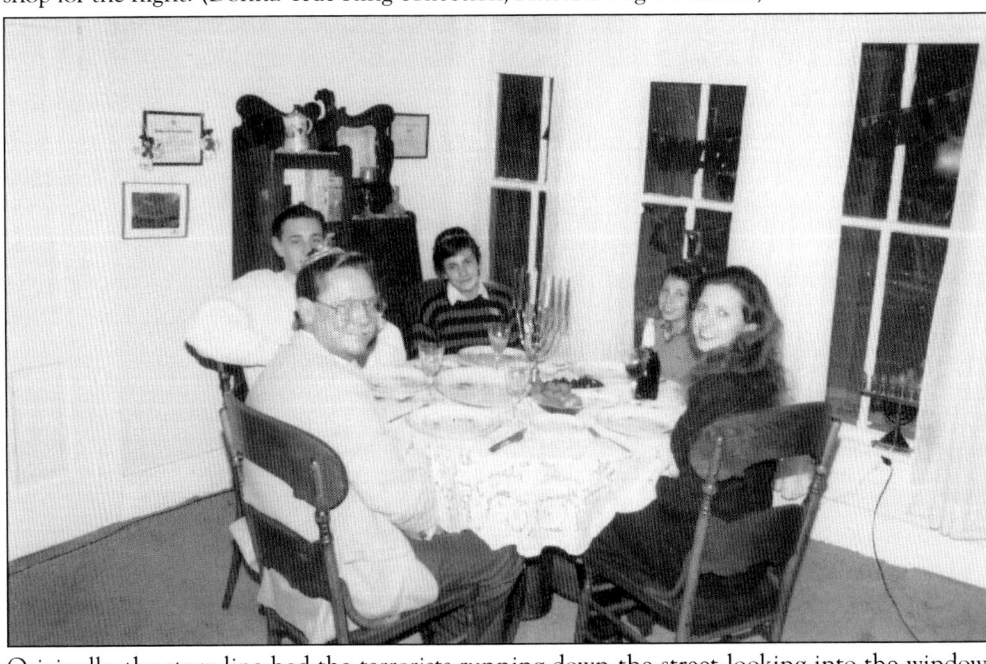

Originally, the story line had the terrorists running down the street looking into the windows of homes until they came to the church. Several residents' homes were decorated for Christmas and Chanukah. To create this scene, Mary Mead Jensen's bedroom was transformed into a dining room. In the end, all the interior shots of people's homes ended up on the cutting-room floor. (Mary Mead Jensen Family collection, Historic Highlandlake.)

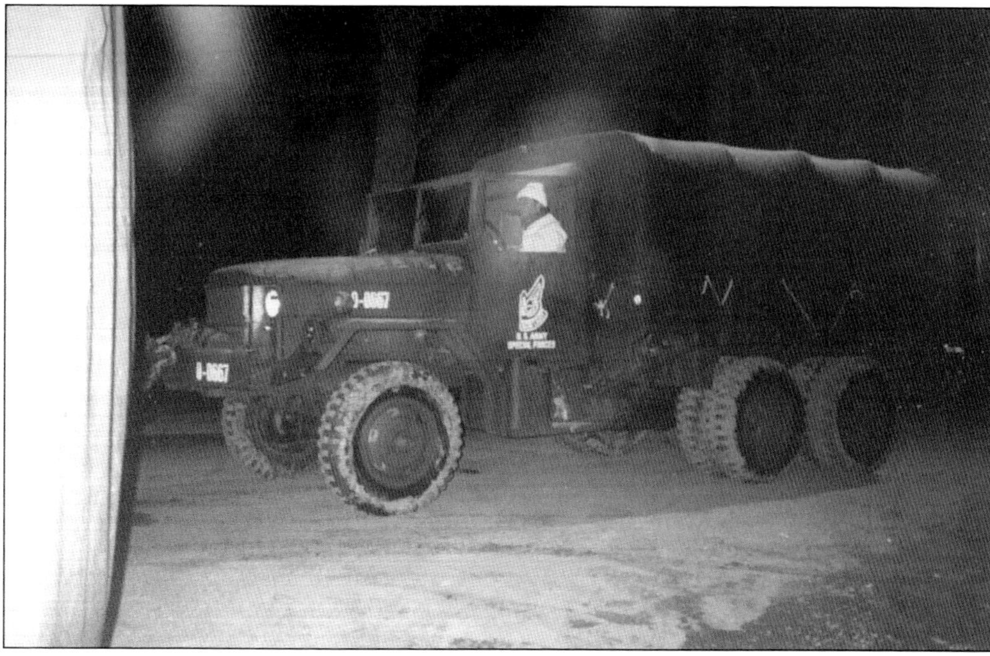

After actor Bruce Willis, alias John McLain, takes out a few bad guys, the Army arrives, and the terrorists make a break for it on snowmobiles. Shown here is one of the Army vehicles used in the film. Other vehicles were painted to look like Virginia State Patrol cars. (Mary Mead Jensen collection, Historic Highlandlake.)

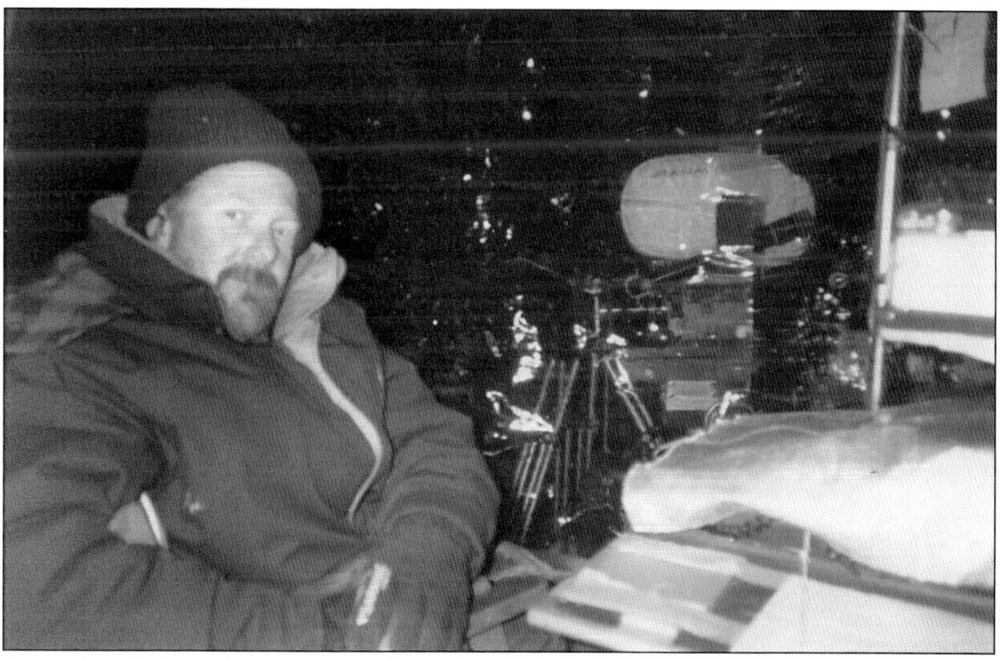

An unidentified soundman poses near his equipment. Since all filming was done at night, with wind chills sometimes reaching 20 degrees below zero, clear plastic shelters helped to protect expensive equipment from the wind and blowing snow. The soundman is so bundled up, it is a wonder that he can even operate his machine. (Mary Mead Jensen family collection, Historic Highlandlake.)

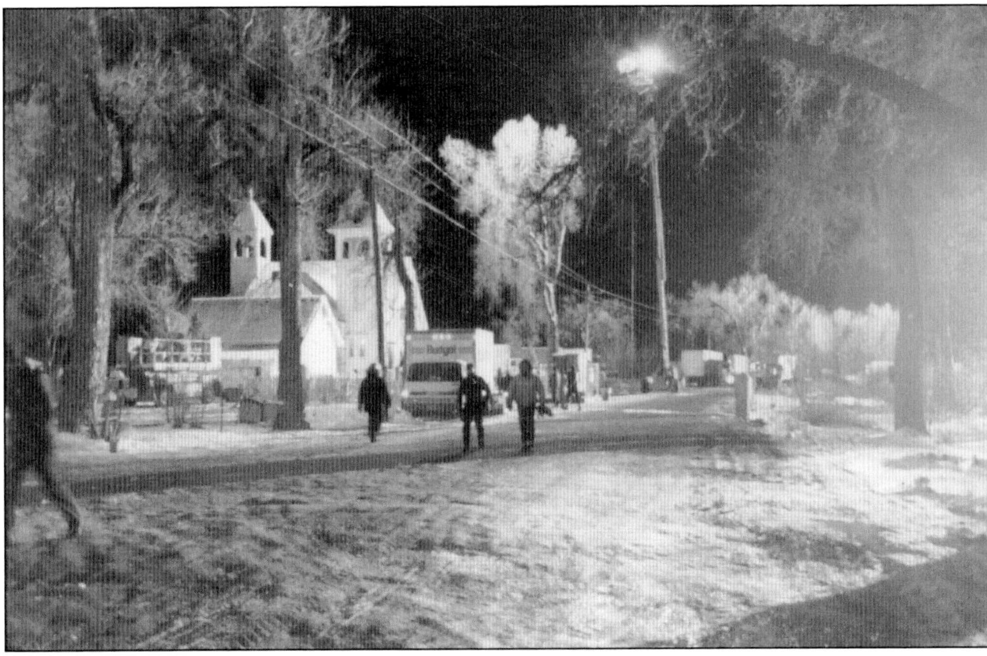

In the photograph above, people are just starting to gather for the night's filming. Rental trucks are parked along the road. To the left of the building on the large boom, a man is barely visible spraying snow onto the roof of the shed. The photograph below shows just how many people were involved in setting up for a shoot. Directly in front of the man standing on a ladder is Finnish American director Renny Harlin. Harlin is seated on a boom that could be raised or lowered as needed. (Both, Mary Mead Jensen family collection, Historic Highlandlake.)

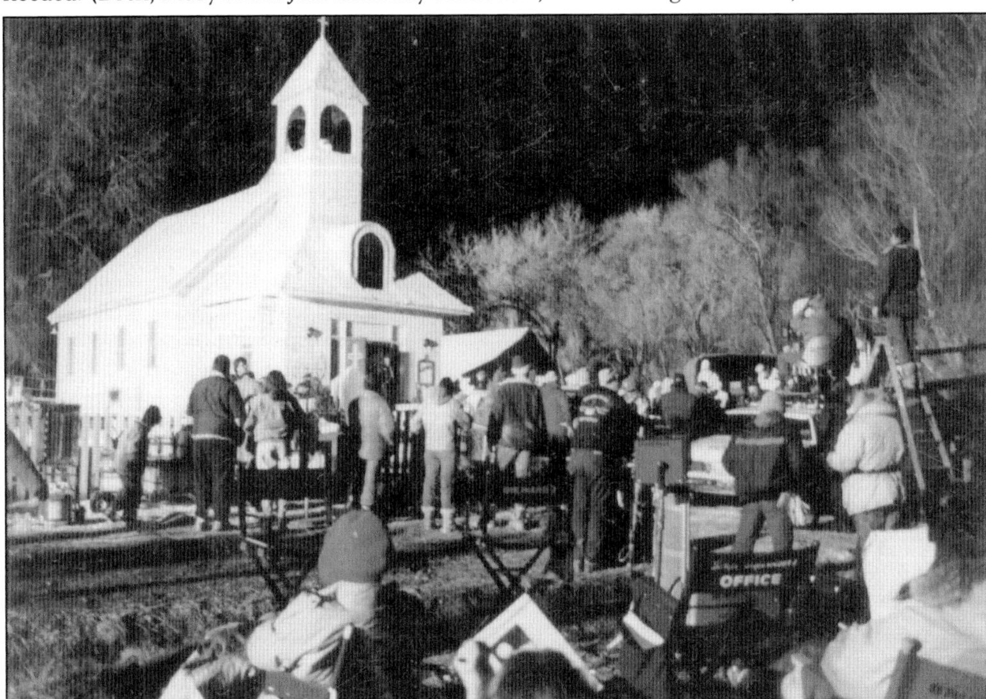

The day before filming started, Bruce Willis flew in on a helicopter to inspect the location. He shook hands with residents, signed autographs, and took one of the snowmobiles out for a spin onto the frozen lake and in the fields to the east and south of the lake. Once filming started, he was in either his trailer or shooting a scene. However, one resident managed to snap two candid photographs. In the photograph at right, he appears to be watching from the sidelines. The photograph below shows him still in costume, walking back to his trailer after shooting a scene. His two bodyguards always accompanied him along the short walk to and from his trailer. (Both, Mary Mead Jensen family collection, Historic Highlandlake.)

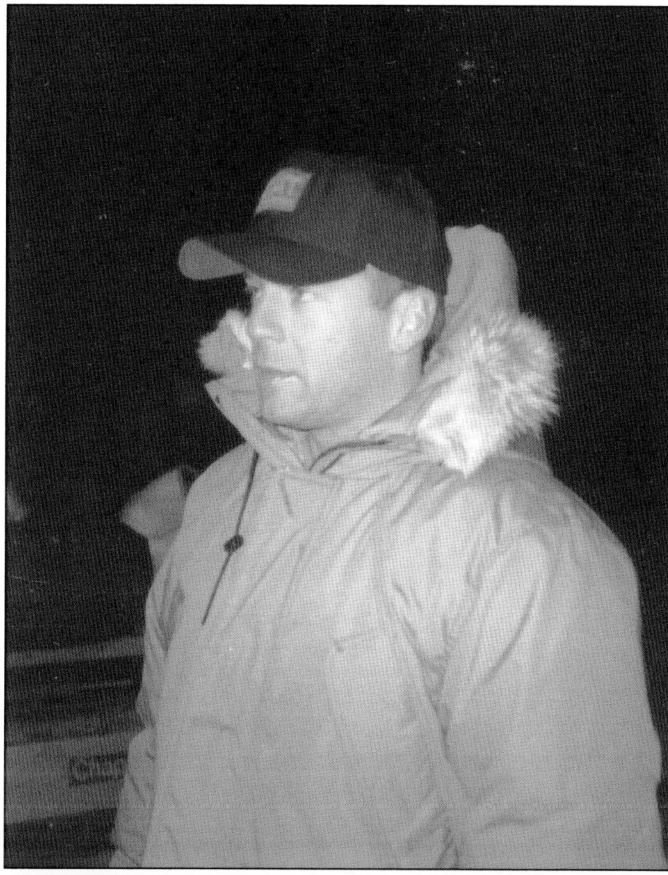

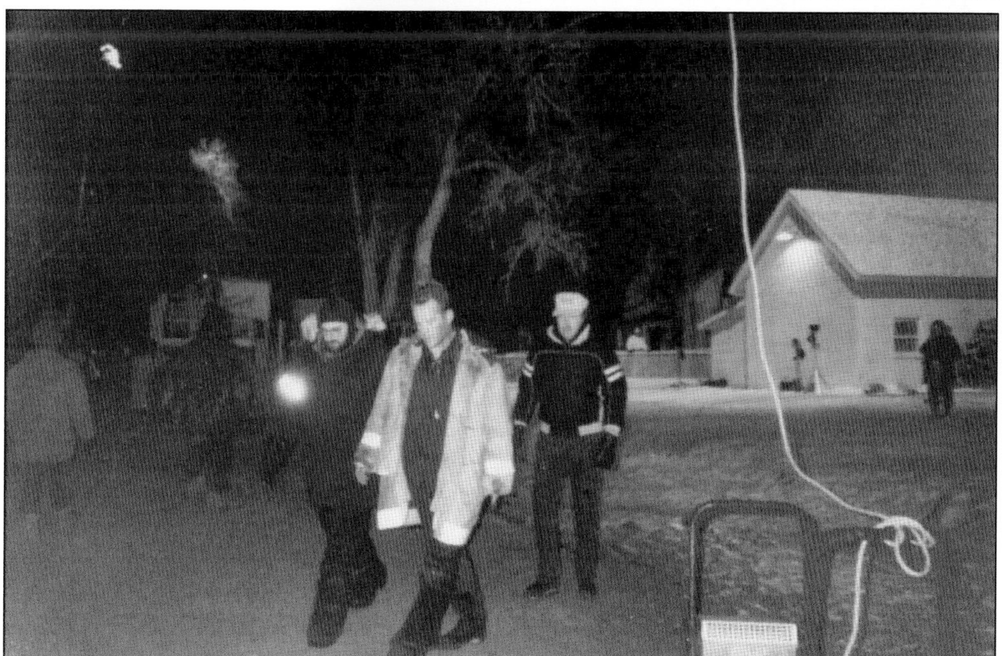

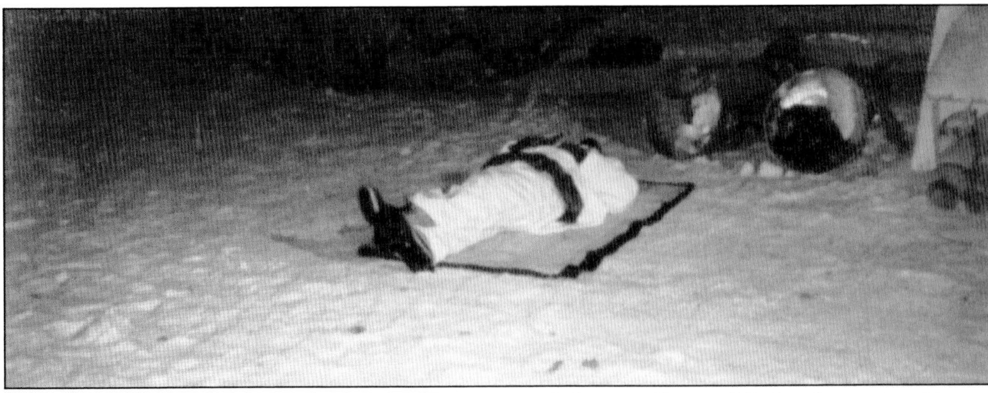

Often, a scene needed to be shot multiple times until director Renny Harlin determined it was just right. Long waits often occurred between shots. In this photograph, a stuntman takes a quick nap while waiting for a call to return to the set. (Mary Mead Jensen family collection, Historic Highlandlake.)

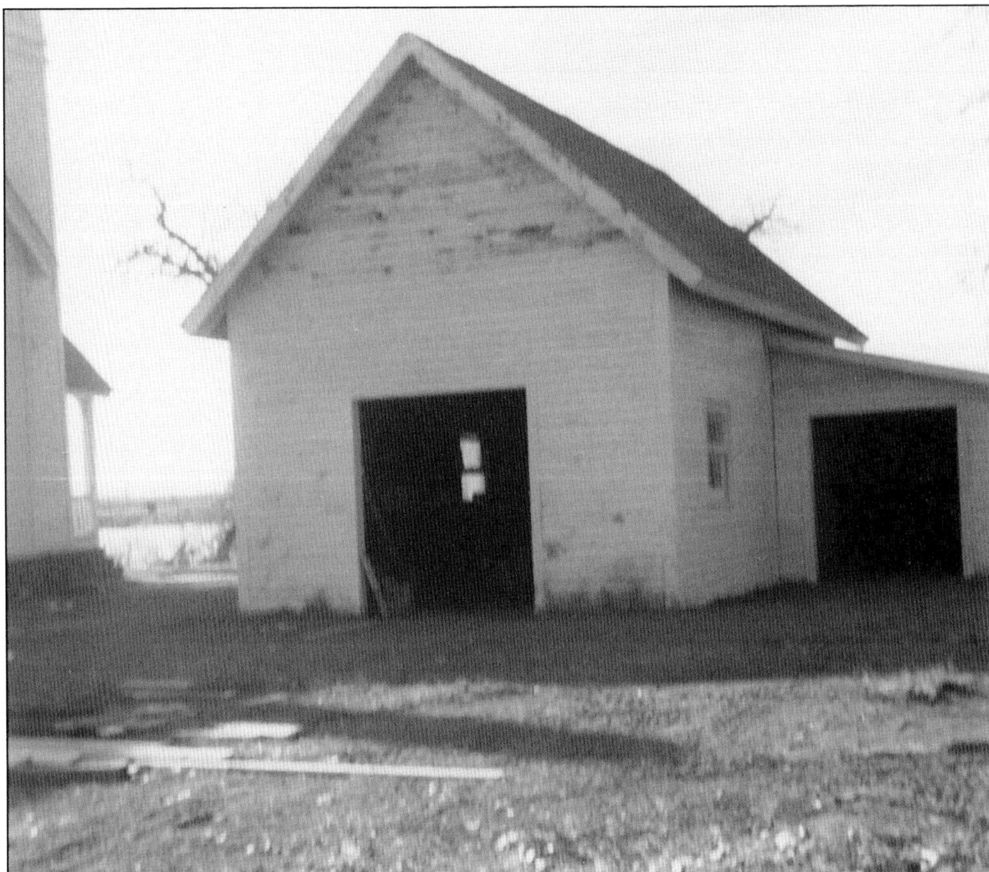

Used mainly for storage, this shed was donated by the studio to the Historic Highlandlake Church Association, now known as Historic Highlandlake. Treasurer Ron Brossman accepted sealed bids, and Longmont realtor Bill Nelson eventually won the right to purchase the building. The money raised went toward repairs and painting the Highlandlake Church. (Mary Mead Jensen family collection, Historic Highlandlake.)

Seven

THE GREAT BLIZZARD OF 1913

The Blizzard of 1913 still sits on the records books as the worst snowstorm ever recorded in Colorado. The snow started falling on Monday, December 1, and except for occasional breaks, did not stop for five days. After that, snow fell off and on until the first week in January 1914. In the first days alone, the Mead area recorded a little over three feet. Denver was hit harder, logging in at 45.7 feet.

Highlandlake pioneer Isadore Terry True kept a daily diary, giving rare glimpses into this difficult time. Below are a few excerpts from her diary. Note that "dad" is R.S. True, Isadore's husband, and Carl and Ren are her sons.

December 1, 1913: Snowing hard.

December 5: Snow falls all day till evening. Am simply shut in. Ren cut a passage thru to the closet (outhouse) and managed to get to the barn before night.

December 8: After an hour or two of hard work, Dad and Ren get way broken around line so can wash and get clothes out. Never in my life saw anything like the depth of snow.

December 11: Eat, read, and sleep the usual routine. Men shovel snow—their daily occupation. Looks as through will be for rest of year at least.

December 14: Lovely day. Carl hitches up and takes boys to Mead to see the rotary open the sugar road. Trains have not yet been able to get north of Mead. Engine off track in Forbes's (Coates) field.

December 19: Snowing when we get up, keeps up till p.m. Seems as though snow never will stop. It blows and drifts again.

December 25: Clear and cold. Take clothes and presents to Carl's, then have nice turkey dinner.

December 27: Take sleigh ride to Mead toward night, get mail and groceries. Many are having sleds and runners made for it looks as though we will have good sleighing until next summer.

January 4: Bright and fair. Some of the church members opened church today.

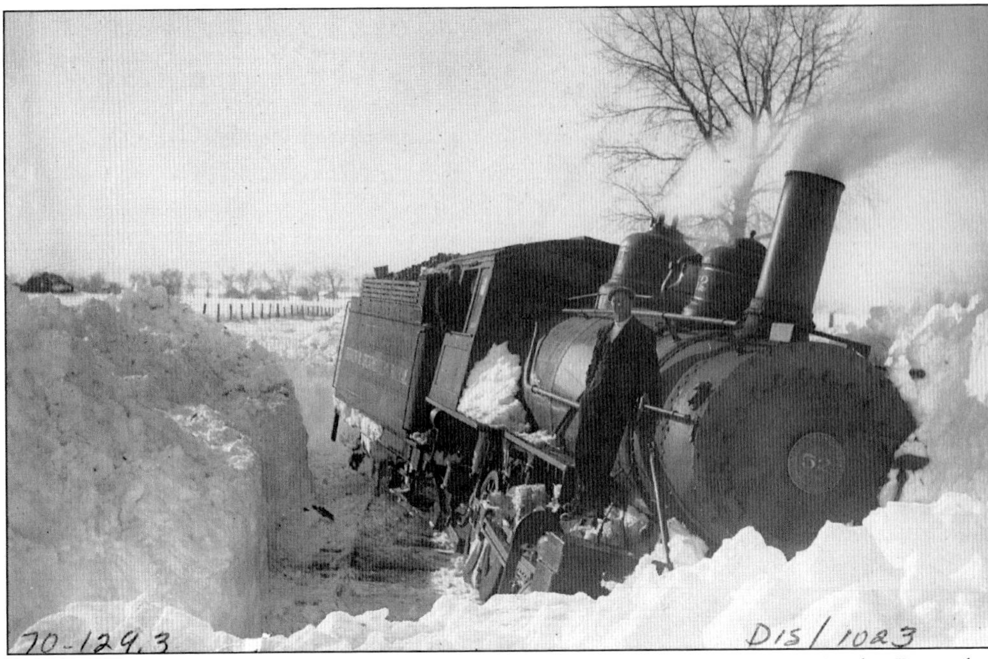

During December 1913, over three feet of snow paralyzed Colorado's Front Range. In this December 14, 1913, photograph, an unidentified engineer for the Great Western Railroad stands watch over Engine No. 52, which derailed in Forbes Coates's pasture. Highlandlake pioneer Isadore Terry True writes in her diary about going to see the marooned engine with her son Carl True and grandsons Clyde and Kenneth True. (Longmont Museum and Cultural Center.)

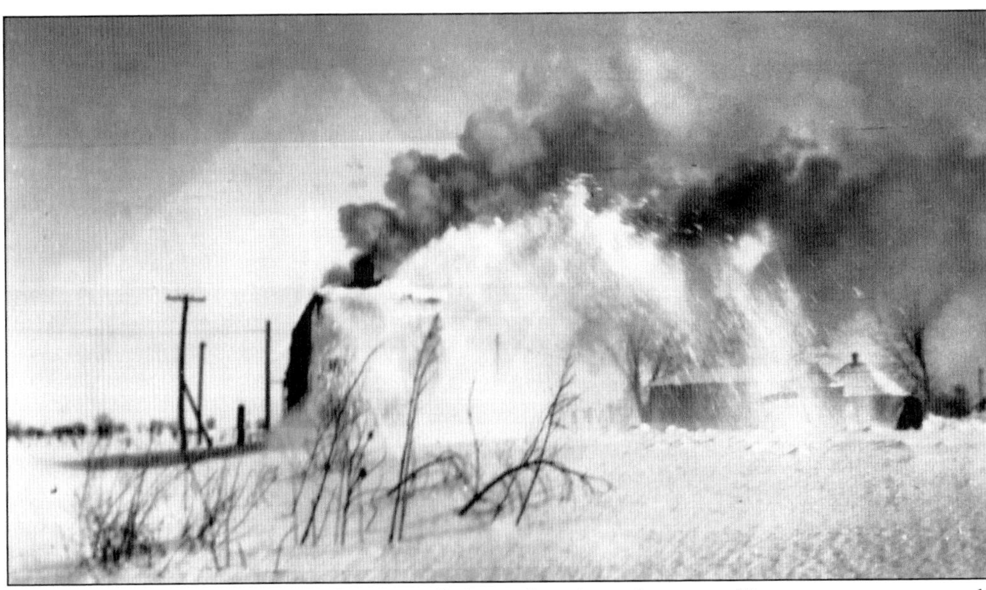

The sight of the rotary plow clearing off the railroad tracks east of Longmont was a ready excuse for Gertrude Spencer, the future wife of Mark Kiteley, to throw on a warm coat, grab her Brownie camera, and snap a photograph to document the event. (Nancy Kiteley collection, Historic Highlandlake.)

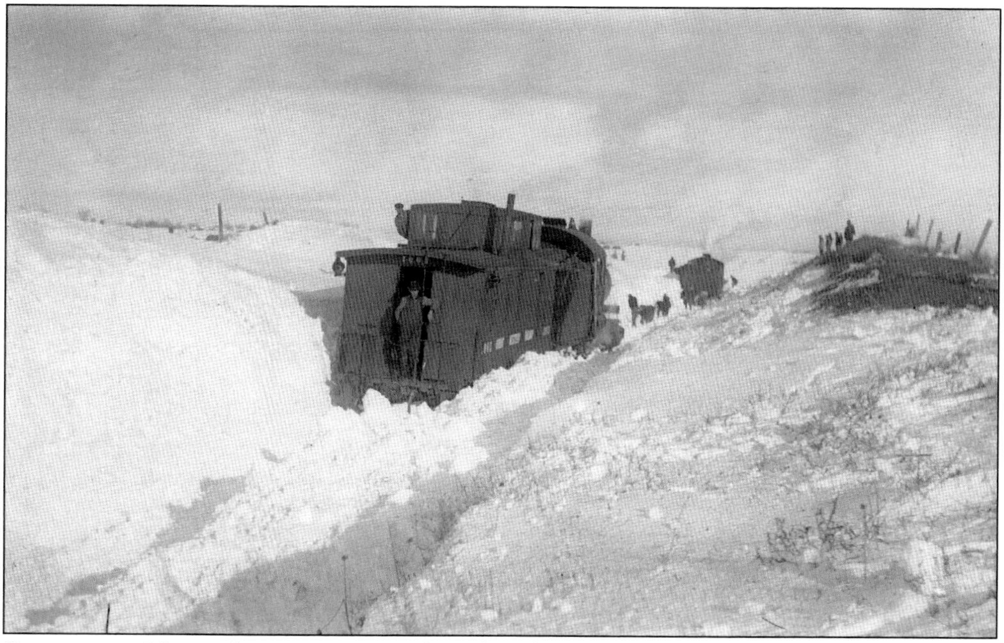

Spectators safely watch on a rise as workers shovel off the tracks. Opportunities abounded for both amateur and professional photographers to document the efforts to clear the way. Two crew members on the caboose seem as interested in the photographer as the photographer is in them. Off in the distance, the community of Highlandlake can be seen. (Longmont Museum and Cultural Center.)

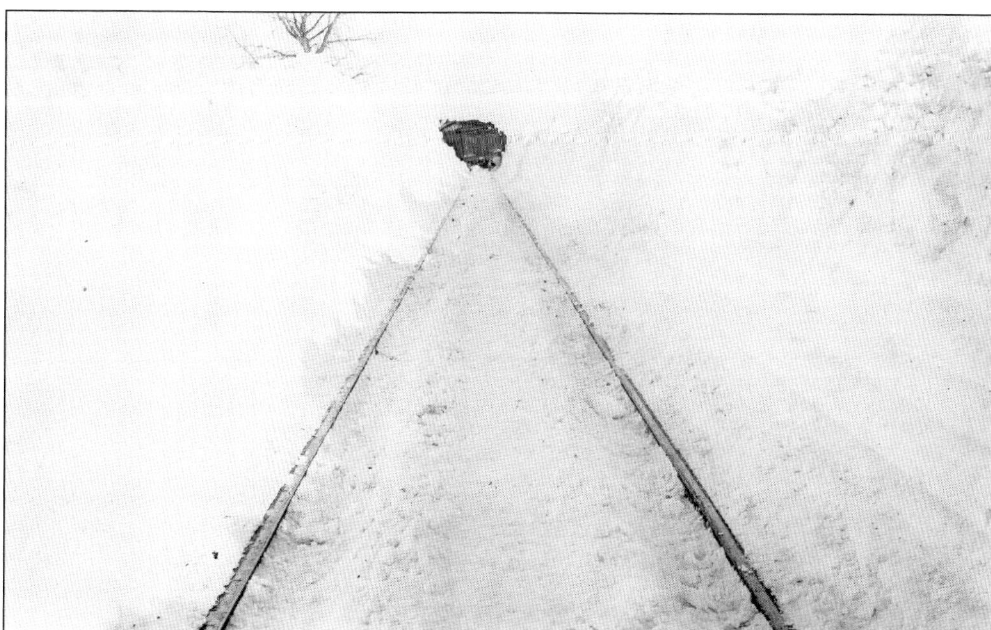

Walls of packed snow are piled high on either side of the tracks. That is Great Western Railway's hapless Engine No. 52, along with an unidentified member of her crew, still waiting patiently for assistance to arrive. (Longmont Museum and Cultural Center.)

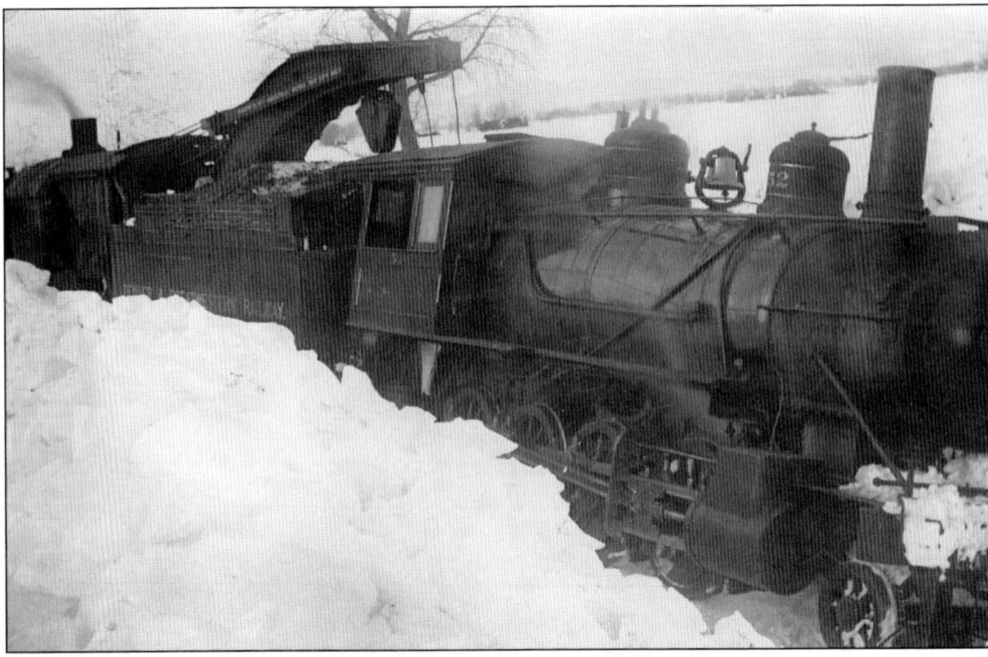

After all other efforts to get Engine No. 52 back on the rails failed, success was finally achieved after the Great Western Railway borrowed Colorado Southern's "Big Hook." Once the engine was righted and moved out of the way, the rotary plow could continue clearing the tracks north of Mead to Johnstown. (Longmont Museum and Cultural Center.)

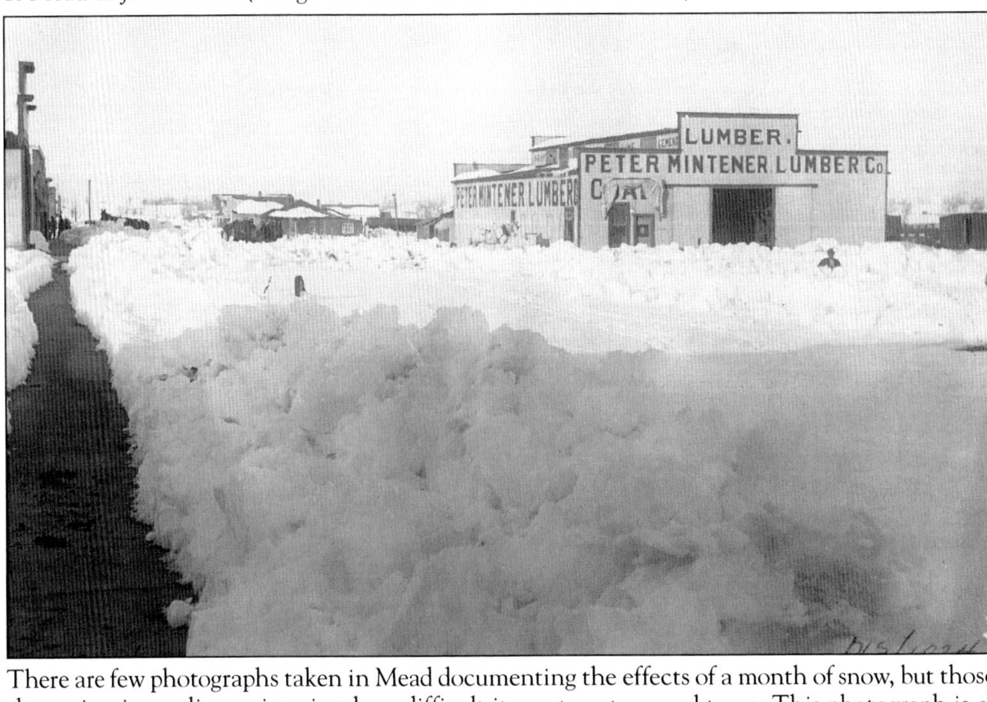

There are few photographs taken in Mead documenting the effects of a month of snow, but those that exist give a glimpse into just how difficult it was to get around town. This photograph is of Fourth Street looking north. On the right is the Peter Mintener Lumber. Directly across the road is the Rose Meeting Hall. (Longmont Museum and Cultural Center.)

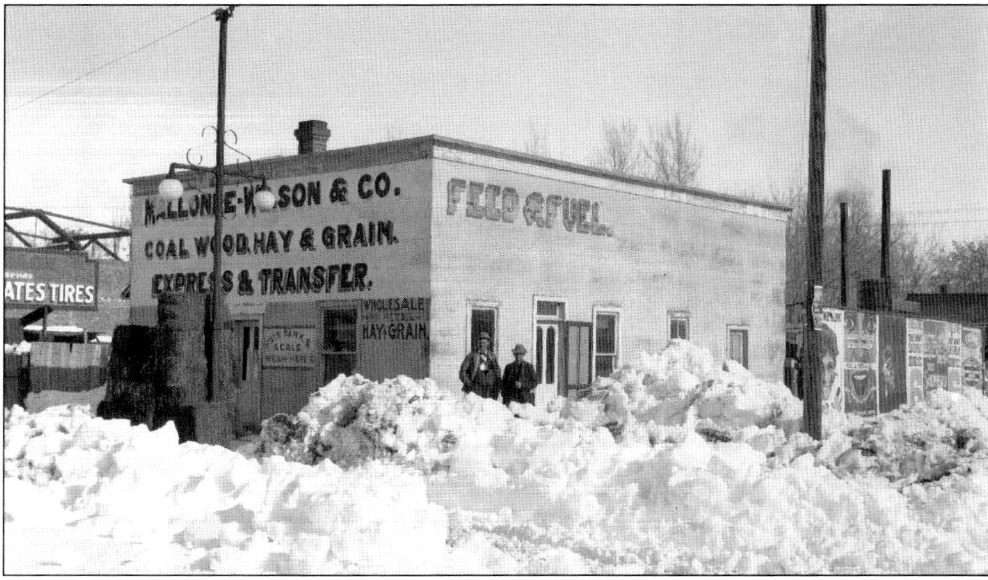

The Malone-Wilson and Company store carried feed, fuel, coal, and wood, and if he wished, a farmer could weigh his load of sugar beets on the store's Fair Banks scales. Along the side of the building are plastered several advertisements for patent medicines and a theater production. This building is no longer standing. (Longmont Museum and Cultural Center.)

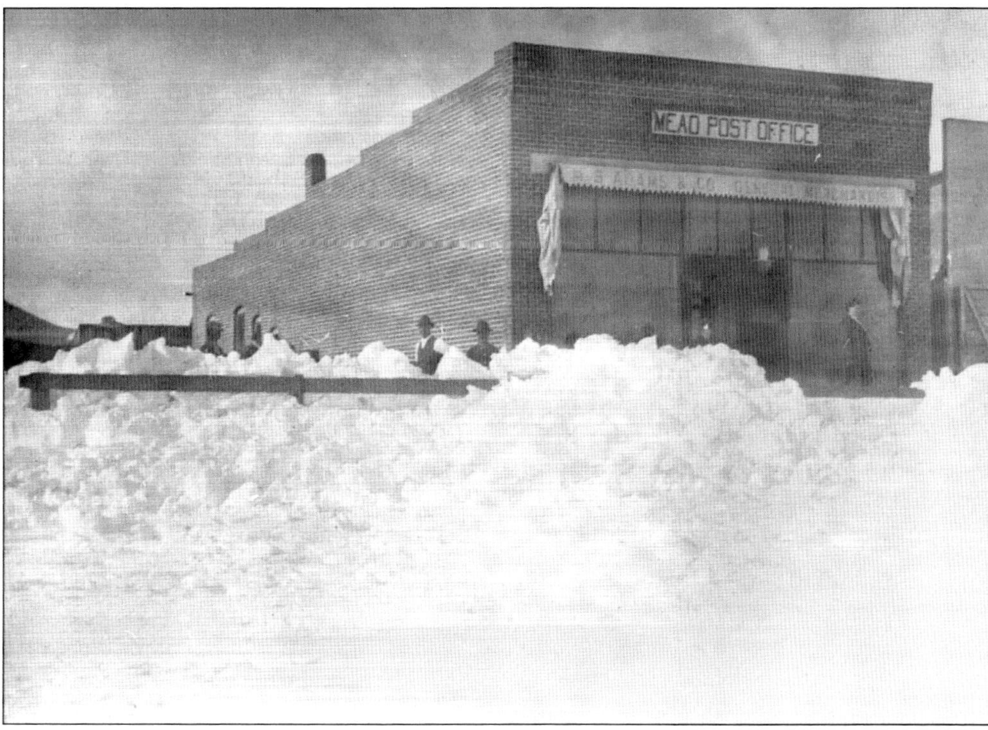

M.S. Adams and Company was owned by Marion Squire and Rebecca Adams. In 1906, they purchased a business lot on the northeast corner of Welker Avenue and the old Main Street. That same year, Marion became Mead's first mayor. Business was so good that they built this large brick store on Fourth Street in 1913. (Longmont Museum and Cultural Center.)